INTRODUCING THE
LATHE

INTRODUCING THE

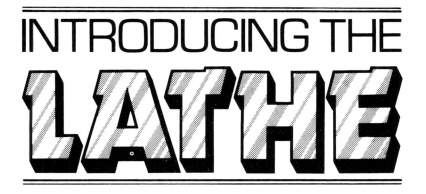

STAN BRAY

Patrick Stephens, Wellingborough

First published in 1984

British Library Cataloguing in Publication Data

*Bray, Stan
PSL model engineering guide.
1: Introducing the lathe
1. Engineering models
I. Title
620'.00228   TA177*

ISBN 0-85059-727-7

*Patrick Stephens Limited is part of the
Thorsons Publishing Group.*

*Photoset in 10 on 11 pt Garamond by Manuset
Limited, Baldock, Herts. Printed in Great Britain
on 100 gsm Fineblade coated cartridge, and bound,
by The Garden City Press, Letchworth, Herts, for
the publishers, Patrick Stephens Limited,
Denington Estate, Wellingborough, Northants,
NN8 2QD, England.*

CONTENTS

	Introducing The Lathe	7
Chapter 1	What to buy	8
Chapter 2	Setting up	20
Chapter 3	Cutting tools	26
Chapter 4	Holding the work	36
Chapter 5	Tool-posts	48
Chapter 6	Turning, drilling and boring	53
Chapter 7	Threading	66
Chapter 8	Milling	74
Chapter 9	Other operations	81
Appendix A	Safety	92
Appendix B	'Do's' and 'dont's'	93
Appendix C	What can be made?	94
	Index	95

Decimal equivalents of fractions of an inch

Fraction in	Decimal in	Fraction in	Decimal in	Fraction in	Decimal in
1/64	.0156	11/32	.3437	43/64	.6718
1/32	.0312	23/64	.3593	11/16	.6875
3/64	.0468	3/8	.3750	45/64	.7031
1/16	.0625	25/64	.3906	23/32	.7187
5/64	.0781	13/32	.4062	47/64	.7343
3/32	.0937	27/64	.4218	3/4	.7500
7/64	.1093	7/16	.4375	49/64	.7656
1/8	.1250	29/64	.4531	25/32	.7812
9/64	.1406	15/32	.4687	51/64	.7968
5/32	.1562	31/64	.4843	13/16	.8125
11/64	.1718	1/2	.5000	53/64	.8281
3/16	.1875	33/64	.5156	27/32	.8437
13/64	.2031	17/32	.5312	55/64	.8593
7/32	.2187	35/64	.5468	7/8	.8750
15/64	.2343	9/16	.5625	57/64	.8906
1/4	.2500	37/64	.5781	29/32	.9062
17/64	.2656	19/32	.5937	59/64	.9218
9/32	.2812	39/64	.6093	15/16	.9375
19/64	.2968	5/8	.6250	61/64	.9531
5/16	.3125	41/64	.6406	31/32	.9687
21/64	.3281	21/32	.6562	63/64	.9843

Conversion — millimetres to inches

mm	in	mm	in	mm	in	mm	in
1	.039	26	1.024	51	2.008	76	2.992
2	.079	27	1.063	52	2.047	77	3.031
3	.118	28	1.102	53	2.087	78	3.071
4	.157	29	1.142	54	2.126	79	3.110
5	.197	30	1.181	55	2.165	80	3.150
6	.236	31	1.220	56	2.205	81	3.189
7	.276	32	1.260	57	2.244	82	3.228
8	.315	33	1.299	58	2.283	83	3.268
9	.354	34	1.339	59	2.323	84	3.307
10	.394	35	1.378	60	2.362	85	3.346
11	.433	36	1.417	61	2.402	86	3.386
12	.472	37	1.457	62	2.441	87	3.425
13	.512	38	1.496	63	2.480	88	3.465
14	.551	39	1.535	64	2.520	89	3.504
15	.591	40	1.575	65	2.559	90	3.543
16	.630	41	1.614	66	2.598	91	3.583
17	.669	42	1.654	67	2.638	92	3.622
18	.709	43	1.693	68	2.677	93	3.661
19	.748	44	1.732	69	2.717	94	3.701
20	.787	45	1.772	70	2.756	95	3.740
21	.827	46	1.811	71	2.795	96	3.780
22	.866	47	1.850	72	2.835	97	3.812
23	.906	48	1.890	73	2.874	98	3.855
24	.945	49	1.929	74	2.913	99	3.888
25	.984	50	1.969	75	2.953	100	3.937

INTRODUCING THE LATHE

There have been many significant inventions or developments in the history of the world, but few can have had so great an effect on industrialised society as has the development of the lathe. At first sight it may appear to be just a machine for making accurate round items but in fact, with its built-in squareness, it has a far wider range of uses than that, and for both the man who likes to tinker and the one who is a serious metalworker, the lathe can be persuaded to tackle a whole range of most unlikely jobs. I use the word 'persuaded' because it is the expertise of the operator, his or her ingenuity, that finally decides just how versatile the machine can really be. Many thousands of lathes have been sold to the amateur over the years and it seems highly probable that many more would like to own one. Some surely think they do not know enough of how to operate the machine, others probably that they cannot afford to purchase such an item.

I hope that I will be able to persuade such doubters that their fears are without foundation. I also hope to be able to offer a little help and advice to the lathe owner who is not too sure of how to get the best from his or her machine. It is surprising how little dodges can make life so much easier. Not all the ideas that I will give are my own, as many have been picked up over the years from other enthusiasts and some of these have been modified, in turn, by my own ideas. Perhaps, with a bit of luck, I will prevent some people from making some of the mistakes that I have made over the years. Perhaps, too, I can offer some guidance on what to buy and the order in which accessories should be bought—important because once you have become an enthusiast, you will want to be constantly adding to your range. There is no need to worry about the price of these because much of what you will want is easily made and making it just adds to the fun.

CHAPTER 1

WHAT TO BUY

Although this chapter is intended as a guide to the purchasing of the lathe, it would be as well to describe first of all the machine and its various parts, as knowledge of the functions of these parts will help when thinking about what to buy. You will obviously have some idea of the use to which your lathe is to be put and this too will have a bearing on what you will need when you come to actually paying out the cash. Another thing to consider is where it will be housed. If it has to be kept in the wardrobe and be taken out as required, then it is probably as well to avoid the purchase of a machine weighing half a ton or so and to stick to one of the miniatures. Equally, if it is intended for construction of half-scale

A simplified drawing of a typical lathe showing the basic parts before assembly of the saddle.

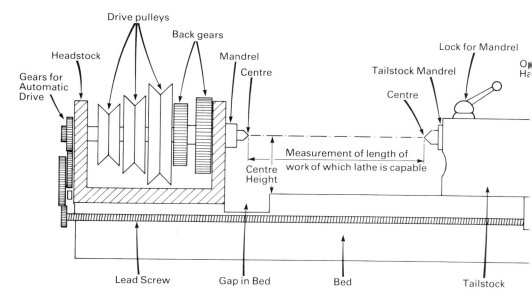

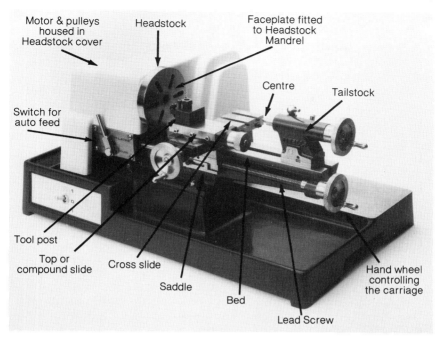

A modern small lathe with a built-in motor. (Photographed by courtesy of Cowells Ltd, Norwich.)

traction engines there is little point in purchasing a watchmaker's lathe. Ultimately, it is probably best to buy as large a machine as possible, because if you do not at some stage find that your machine is not large enough for your requirements then you are a most unusual individual!

At the basic level, it is fair to say that all lathes consist of virtually the same parts. They may differ in shape a little but they are all essentially the same. The first consideration is size; two measurements are usually quoted by manufacturers, namely the centre height, which is roughly half the diameter of the work that can be turned, and the length of the work that can be held. Some manufacturers make lathes of the same centre height but of different lengths and here, again, it is worth trying to establish your requirements. I have often felt the need for greater centre height but never for a greater length. However, if my hobby was making fancy walking sticks out of metal, then I have no doubt whatever that the extra length would be more important than the extra height. If you can only afford a small lathe or if you only have room for such a machine, and the machine is for hobby purposes, then do not despair. A lot of fun can be had and fine work done on the small lathes. I happen to be lucky enough to have a small lathe and a large lathe, and I get equal fun from both.

THE LATHE DESCRIBED

A lathe consists of a main body, known as the 'bed'. This can be made with a flat top or it may have one or two vee-shaped pieces which act as guides, or it may even be round. In any event, it will, except in one or two very cheap lathes be of cast iron. The length of the bed will be the determining factor as to how long a piece of metal can be worked on. At one end of the bed (the left, unless you buy a left-hand lathe, and there are such things) will be mounted assorted hardware, including the part that revolves. The overall name for this is the 'headstock'. In some cases it is bolted on and in others cast integral with the bed. Housed in the headstock will be the mandrel, the revolving part. It will be set in a bearing of some sort—either some form of ball or roller race or a plain metal bearing. Some lathes are arranged so that these bearings can be adjusted, though most sold for amateur use have no such feature. The mandrel itself will be of steel and will have on it: the pulleys which are used to make it revolve; a means of holding a chuck or faceplate; and, usually on the back, a gear wheel. The mandrel should be as large as possible and one item that should be noted is the size of the hole pierced through it. This will, to a large extent, govern what can be done in the way of turning. Some lathes do not have a hollow mandrel and it is a facility that will be sadly missed when the lathe is in use. Sometimes just in front of the headstock, the lathe bed is cut away and this is called a 'gap' bed. The reason is obvious—a greater size of work can be accommodated. On the more expensive lathes, a piece of the bed is made removable thus giving the best of both worlds. It is possible that behind the front mandrel bearing there will be a small cluster of gears. These are known as the back gears, in spite of being in the front, and are a most useful feature for turning heavy work. On the smaller lathes the same idea is often used but, instead of through gears, the effect is obtained by means of a belt-changing system.

Before leaving this part of the lathe we should come back to the mandrel and at the end where the chucks and things will go, will be a taper, on the inside. This is a means of fixing tools by friction, and is highly effective. It will invariably be what is know as a morse taper, and is numbered zero or one on the small machines and two or three on the larger. The larger the taper, the better the grip on the tool being held by it, but the taper size is of necessity governed by the diameter of the mandrel.

The next part of the lathe to be considered is the carriage or saddle. It is so named because it straddles the bed like a saddle and moves along it like a carriage. Upon it there is a cross-slide to allow tools to be moved across the bed and sometimes there is a top-slide which assists in the turning of small tapers and radii. Not all lathes are fitted with top-slides, particularly in the smaller models. The carriage itself is moved along the bed by being wound along a lead screw. Though usually found at the front of the bed, it is sometimes at the back; also on square- and round-bed lathes it may run under the bed. Movement is controlled by a hand-wheel which is either attached to the carriage or is at the end of the bed. On all lathes of reasonable quality, the carriage can be made to operate under its own power. The cross-slide is, as has already been stated, a means of taking tools (or work) across the bed and on very high quality lathes this too will have a means of power operation. Otherwise it is wound across using a handle or hand-wheel. Some manufacturers offer an extra long cross-slide as an alternative and this is well worth consideration as it can greatly increase the scope of the work of the lathe. It is also desirable that the cross-slide should be as wide as possible. The top-slide, sometimes called a compound slide, looks like a small cross-slide. It will almost certainly have a means of turning tapers by turning the slide round and will contain the means of holding the tool-posts, the various types of which will be discussed later. The cross-slide and possibly the top-slide will have in them a series of slots called, for obvious reasons when examined, 'tee' slots and these are used for securing items to the slide.

At the end of the bed opposite to the headstock, we will find an item called the tailstock. This is used for supporting work at its extremities, for holding tools of various kinds and for drilling holes. The tailstock consists of a casting with a mandrel in it, connected to a hand-wheel or, less frequently, to a lever, by means of which the mandrel can be moved in and out of the casting. There will be some means of locking the tailstock mandrel to prevent vibration working it in or out and there will be a lever to lock the tailstock to the bed to prevent the casting moving as a whole when under pressure. The mandrel will again have an internal taper which will almost certainly be a morse of a size dependent on the lathe. There is usually some form of gradation marking to enable one to know how far in or out the mandrel has been moved. These gradation marks will either be on the mandrel or on the wheel. Similar gradations will appear on the hand-wheels used for traversing the top-slide, cross-slide and, possibly, the carriage. These items altogether make up the basic lathe.

I have already pointed out that when purchasing a lathe it must very much be an individual choice for the person concerned, dependent on size and money available. It is as well when purchasing new to obtain details of various lathes and sit and study them. Find out what each lathe will do and what you will get for your money. Many lathes come as a basic unit, with the only other items supplied being a couple of centres, (about which more later) and a faceplate. There is frequently no motor and no chuck, and whilst it is quite possible to use a second-hand washing-machine motor, it is as well to know what you are getting for your money. The modern trend, however, is to fit a motor as an integral part and possibly supply a chuck of one sort or another. This seems a trend for the better, but is an integral motor such a good thing?. What if, after a year or two, the motor breaks down, will it be possible to replace it?. The answer to this question needs very careful consideration and it may be that, in the long run, a separate motor is the most economical.

ACCESSORIES

Assuming we have decided on our lathe, then the next question to ask is what else do we need. The plain answer is not a lot, if we are willing to use a great deal of patience and ingenuity, but then again it could be that, rather than make lathe bits, the cash is available to buy them so that model making can start straight away. Below are the accessories that are available, listed in the order of what I regard as priority of purchase:

Tailstock drilling chuck
Four-jaw chuck (useful for most work)
Three-jaw chuck (quick to use, but very limited)
Vertical slide (useful for simple milling operations)
Travelling 'steady' (used to support long work)
Revolving centre
Tailstock die holder
Rear tool-post (useful in multiple operations)

Above right *The EMCO Compact 5 lathe. Slightly larger than the Unimat 3, it is a useful lathe for those with only a small amount of room but who wish to tackle something larger. Shown here with drilling attachment.*

Below *A Unimat 3 lathe, a small lathe available with a wide range of accessories.*

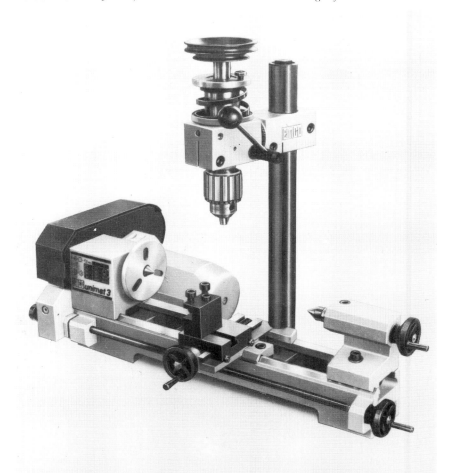

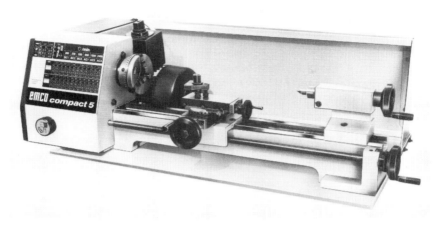

Below *The Maximat Super 11, shown here with the milling attachment is from the same manufacturer as the Unimat 3 and the Compact 5 but is suitable for much larger work. It is available with its own stand.* (All the photographs of this range are by courtesy of E.M.E. Ltd.)

BUYING SECOND-HAND

If a new lathe cannot be purchased then a second-hand one will have to be considered. There are some very good second-hand machine dealers and although their lathes tend to be a little higher in price than machines sold on the open market, if you are not too sure of youself it may be as well to try one. Examine the lathe carefully whether you buy privately or from a dealer. Look at the general condition and do not be fooled by new paintwork. Take with you a piece of 25mm by 50mm wood about 60cm long. Put it under the nose of the mandrel and lever the mandrel up. The wood cannot damage the lathe but if bearings are worn then any wear will show in the form of movement. If there is wear it may be as well to contact the lathe's manufacturer to see what repairs will cost. Run the carriage up and down—it will almost certainly be easier to operate at the part near the mandrel nose than at the end. The question is how much? Such wear must not be excessive. Look at the bed to see how badly it is marked with digs and grooves. If there are some signs of heavy marking, the machine has probably been used badly and it may be as well to look elsewhere for one. Check that the tailstock tightens up correctly and check with your piece of wood for wear on the mandrel. Examine the back gears and the gear-cutting gears, if fitted, for signs of heavy wear. Check all slides for movement, but remember that there is some adjustment as will be described in due course, so do not discard a lathe out of hand for this reason. If an integral motor is fitted, see it running and check the bearings if possible. If chucks are fitted then see how much wear there is on the jaws; badly worn chucks are a waste of time and money and it may be that the price is set too high simply because such useless chucks are included. A most important check concerns the bed itself. It is quite

The Peatol Micro Lathe. Probably the cheapest lathe on the market, it is made of extruded alloy. It comes with motor and chuck, and in spite of its restricted size and unusual construction is a useful little machine for those of limited means.

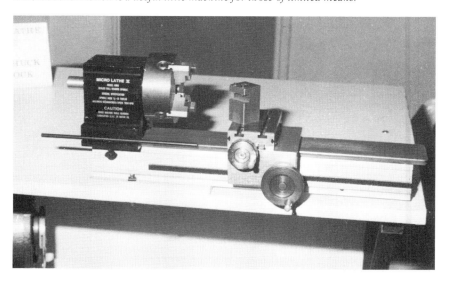

The Peatol Micro Lathe has a lever-feed tailstock, instead of the more usual hand-wheel. This arrangement is available on some other lathes.

possible for the lathe beds to twist and thus become inaccurate. This is not an easy fault to check and it may be that the only way to do so will be to take off the tailstock and look along the bed for any twists. Two pieces of steel bar or two pieces of wood laid across the bed near the ends will assist in this, as it can be seen fairly quickly if they do not line up (both pieces must be the same size, of course). If the owner will allow a test piece to be turned, then a small piece turned off the longest available length will quickly establish whether there is a twist. If the measurement is consistent on the diameter over the whole length then there is no problem whatever.

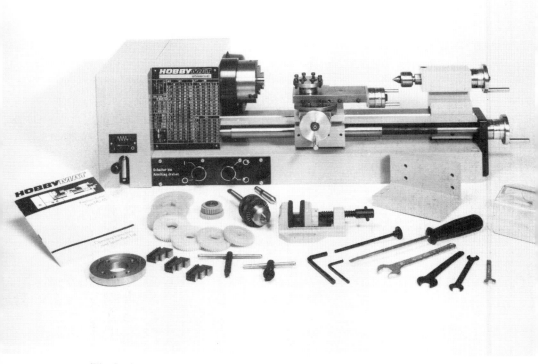

The Hobbymat lathe, a medium size lathe from Eastern Europe. With a 75mm centre height and 300mm between centres it is a similar size to the Compact 5. (Photograph by courtesy of C.Z. Scientific Instruments Ltd.)

It is difficult to explain in full everything to look for in the purchase of the lathe but I hope that at least some of the advice given above will help. Having become the proud possessor of this super machine, we must now learn how to set it up and how to use it. Finally, a word to the person who thinks he or she cannot afford to get a lathe. Do not spurn your local scrap yard—you may not get the Rolls Royce that you would like, but a lathe in any condition is better that no lathe at all. It will soon be possible to correct many of its faults and the fun you get from it will be as great as that of the person who has bought a brand-new super model with all the accessories. You must make sure though that it is not seized up solid—as long as it turns and the carriage traverses, buy it. There is little use though in looking in scrap yards that only deal in motor vehicles. A good general scrap merchant, if possible one dealing in old machinery, is the man to try. The yards of those who deal with factories disposing of lathes which are of no further use to them can often be spotted by the metal swarf lying around—this is a sign that they deal with factories that engage in the type of engineering in which we are interested.

Above *The Harrison M250. Harrisons have only recently entered the hobby market, having previously catered purely for industry. This means that a great deal of engineering expertise has been brought to bear on this machine.* (Photograph by courtesy of T.S. Harrison and Sons Ltd.)

Below *A household name in model-engineering lathes and accessories, Myford have now produced this model 254 S which incorporates many modern features whilst maintaining its suitability for the model market.*

The latest offering from Unimat is a computerised Compact 5. Computerisation gives ease of working and, properly used, can increase accuracy.

BUYING NEW

For the purchase of new lathes it is probably as well to go to a major tool dealer or direct to a manufacturer or importer. Some names have become household words amongst hobbyists using lathes, such as Myford of Nottingham, who are the oldest established supplier of modellers' lathes. E.M.E. Ltd of Willesden, London, import machines from Europe and can supply a wide range of different size lathes. Most recent entrants into the market are firms like CZ Scientific Instruments of Boreham Wood, Herts, with the Hobbymat, and Eric H. Bernfield of Potters Bar, Herts, with the Toyo ML1. There are many other makes available at a wide variety of prices and, in recent years, manufacturers who previously catered only for the professional market have started to introduce lathes for the amateur—of these the best known is the M250 by T.S. Harrison and Sons of Heckmondwyke in Yorkshire. Quite a few lathes are now imported from the Far East and they closely resemble models made in this country and Europe. For second-hand lathes it is as well to scan advertisements in modelling magazines or to go to a dealer. In the small range it is possible to obtain Unimats or Cowells which are both excellent lathes for the beginner. In the larger sizes there are some Myford ML7 models and a few of the Super Sevens. There are plenty of suitable lathes available from lesser known manufacturers, such as Tyzack, and if slightly larger models of, for example five inch centre height are required, then it is possible to obtain them from Boxfords, Colchesters, Harrisons or other reputable manufacturers.

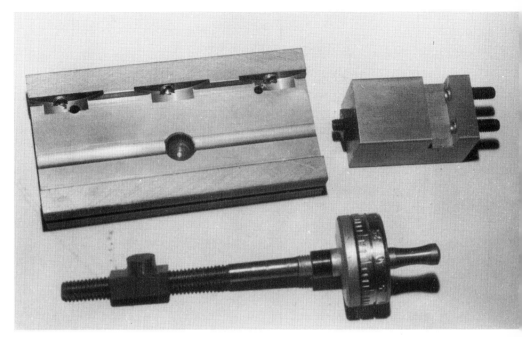

The construction of a cross-slide. The plate on the top left locates on dovetails on the saddle. The nut on the drive screw at bottom right locates in the hole in the centre. On the right is a simple tool-post.

CHAPTER 2

SETTING UP

After parting with your hard-earned cash, you will, if you are like me, be hardly able to contain yourself, as you want to get cracking on the machine. If you have purchased one of the small lathes there is no problem; if the shopkeeper likes the look of your face or you have a good Access rating, you will no doubt be able to bring the machine home with you. After quickly attaching a suitable plug, you will be able to set it up on the kitchen table, put an old nail in the chuck and turn it quickly down and make a panel pin. If the machine is going into a workshop then it can quickly be put on the bench, with just a couple of screws to hold it down. Such machines need little setting up, the beds are rarely long enough to warp and, as long as the surface on which they are to stand is fairly flat and level, there should be no problem. Some, such as the Toyo ML1, even come complete with a neat wooden base that ensures the machine is accurately set up.

If however you have invested in a machine with 3½in or larger centres, then there will be considerably more difficulty in getting going. You will probably have to wait for delivery and, when it is delivered, there may be a certain amount of assembly to be carried out. It is, then, as well to prepare a bench to put it on and, even if you have purchased a lathe complete with an industrial type stand, you will still need to ensure that a strong and level floor is available for it to stand on. Such stands, although somewhat expensive, are the ideal answer to the mounting problem, as not only do they provide a good mounting surface, being designed for the job, but storage space is available as well.

THE WORKSHOP

If a bench is to be constructed then it is worth giving its design a considerable amount of thought. The height is important; it is best to make the bench-top a little under waist high. Not so long ago I had to set my lathe up in a new workshop and I thought that, with rapidly advancing senility, it would be a good idea to make the bench a little higher so that I did not have to lean over. The idea was disastrous and has now been rectified. It proved to be very uncomfortable each time that I wanted to do something at the back of the lathe as I had to stand on tiptoe. One should also take into account what space is needed to accommodate the lathe. If it is put too close to a wall then it will be impossible to get lengths of material through the mandrel. If the lathe is to be used for milling, as so many lathes are, then make sure that there is room behind it for any long pieces of metal that might need to be milled. At one stage, I had my lathe set up by an opening window. The lathe was, amongst other things, used for milling out 5-in-gauge locomotive frames, and these used to be mounted on the cross-slide and extended through the open window, whilst work was being carried out on them. If I had not had that window there, they could not have been machined. It was no accident that the lathe was placed there but the result of earlier experience.

The mention of windows leads naturally to the question of light. There is no substitute for daylight, so an adjacent window is an asset for this reason too. There is of course always the danger of the work coming out of the chuck and going through the window, but with experience this possibility lessens. If we are thinking about light then perhaps we should also think about power. A power point should also be closely adjacent, as long extension leads and trailing flexes are not to be encouraged. A double power point is the best bet, as power will also be needed for light and sooner or later a bench-mounted lamp may be required. Incidentally, in industry, such lamps are, for safety reasons, of low voltage and it is wise to get such a lamp for the home workshop if you can and to wire up the transformer away from the lathe. The other alternative is for a spotlight to be mounted on the ceiling and there are plenty of domestic fittings available that are suitable for this purpose.

THE BENCH

The bench itself must be of stout construction and, if possible, should be bolted to a wall to increase rigidity. It must of course be absolutely level as, if a lathe is bolted to a bench that is either out of level or of uneven surface, problems will **arise with the lathe-bed warping and the machine not being true.** Ideally the bench-top should be of moderately thick metal, though in practice this is rarely possible unless a drip tray or suds tray is purchased. The next best thing is to put metal plates, considerably larger in area than the lathe feet, underneath those feet and to bolt through these. The top, whatever it is made of, must be stout enough to support the lathe for many years and must be a type of wood, if wood is used, that will not warp. Shuttering plywood which can sometimes be obtained from builders' merchants second-hand is an ideal material, as is blockboard, or even a couple of layers of thick chipboard. Whatever is used, it should be covered with a hard plastic sheeting such as Formica or Melamine before the lathe is mounted on it. There will be great deal of oil and cutting fluid around and if it is allowed directly on to the wood it will soon cause trouble.

The bench frame should if possible be made of metal. Welded angle-iron is ideal, but if it is not available one of the slotted angles such as Dexion should be considered. If slotted angle is used then it should be braced diagonally. Cross-braces at the top should be put in such a position that it is these that support the machine rather than the bench top. If wood has to be used tor the frame then make sure it is of a section that will have enough strength. If possible employ second-hand timber that has been well used and is unlikely to warp. New wood almost certainly will give problems before very long and should be avoided. Shelving underneath the bench will add to its strength and provide useful storage

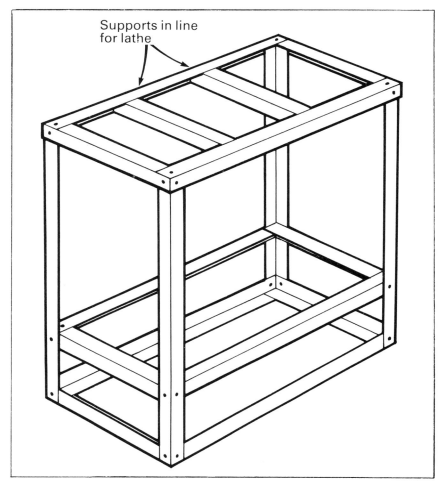

Supports in line
for lathe

Construction of an angle-iron stand suitable for a lathe. The top is covered with steel plate or heavy wood. The two cross-pieces at the top give the lathe its main support. If slotted angle is used, then diagonal braces will be needed at all corners.

space for tools and metal; if you intend to sit down on the job at any time then incorporate a foot-rest in the bench. There is nothing more tiring than having your legs dangling in mid-air for a couple of hours.

Once the bench has been constructed, the next job is to mount the lathe on to it. Here again a great deal of care pays dividends later on. The bed must be level both lengthwise and across. A spirit level of the type used by builders is nowhere near good enough for this purpose and, unless a proper engineer's spirit level is available, then alternatives must be found. There is on the market, as part of a combination set, a device with a dial and needle that will measure any angle very accurately. This used to be available quite cheaply as an individual item, as

A simple, easily made, level indicator. It consists of a free-running pointer on a steel plate. The pointer lines up with a centre punch mark. Accuracy in construction is essential. A plastic protractor screwed to the base would enable graduations to be read.

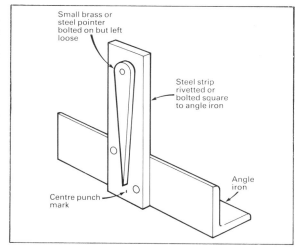

Small brass or steel pointer bolted on but left loose

Steel strip rivetted or bolted square to angle iron

Angle iron

Centre punch mark

A piece of steel plate, with a small-diameter hole right through, will enable centres to be lined up. Even the slightest difference in height will become obvious when the centres are brought to the hole.

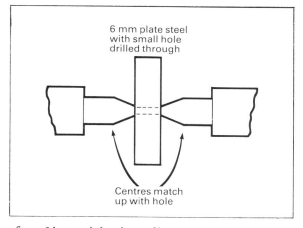

6 mm plate steel with small hole drilled through

Centres match up with hole

shown on page 24, but, as far as I know, it has been discontinued. If such a tool cannot be acquired then take heart, for all is not lost. Half an hour's work will produce such an accurate level that you will wonder how you ever managed without it. Just bolt a piece of flat steel to a piece of angle-iron. Drill a hole in the centre of the steel and put a centre punch mark in line with it. Make up a small pointer that will just hang loosely on a nut and bolt in the hole and the point will line up with the centre punch mark if the angle-iron is placed on a level surface. The longer the angle-iron, within reason, the better and, of course, you must be accurate in your marking out and drilling, as the result depends on this. This simple little tool will be useful for quite a few jobs and if you want to make it more versatile then stick a plastic protractor on the plate and the pointer will read off degrees as well. You will probably need to put a washer under the pointer if the protractor is used.

    The only thing left to do now before the lathe can be used (assuming the motor and other pieces have been assembled) is to see that the centres line up.

Above *A level and angle indicator. This simple device is very accurate. Once available from a number of manufacturers it is believed that it can now only be obtained as part of a combination set, ie, a rule with a square and forty-five degree attachment.*

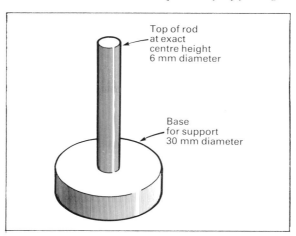

Top of rod
at exact
centre height
6 mm diameter

Base
for support
30 mm diameter

Left *A simple and permanent centre-height finder consisting of a piece of round bar for the base with a smaller pillar set in it. The pillar is at exact centre height enabling the tool to be quickly and easily set to the correct height.*

We shall discuss centres later in the book but for now take it that they are the pointed ends on morse tapers that go in the headstock and tailstock. They should be perfectly in line both horizontally and laterally. It does not matter if they are out laterally as the tailstock almost certainly will have an adjuster, which will enable the centre to be shifted to correct this. It is most unlikely to be out horizontally, but there have been instances of such happenings on lathes manufactured in the Far East and the only way to deal with this problem, as there is no adjustment possible, is to have the tailstock machined if it is too high. To see if the centres are lining up do not rely on the naked eye. Either use a strong magnifying glass, if the check is to be visual, or else drill a small hole through a piece of steel plate about 6mm thick so that each centre will go in slightly less than half way, and after putting it on the headstock centre, wind in the tailstock. Any discrepancy will be immediately obvious.

There is one other thing that should be checked at this stage and it will save a great deal of work later on. All cutting tools, as we shall see, must be set at centre height and it is as well at this stage to make some form of gauge so that it is not necessary each time to put a centre in the lathe, in order to set the tool. A simple block with a short rod sticking out and terminating at centre height is ideal and easy enough to make. Alternatively, scribe a mark on a try square, commonly known as a set square, with a tool that has already been set to the right height. It is then just a case of lining the tool up with the mark. An easy way to set up the tool to an approximate height before final adjustment, if the three-jaw chuck is in use, is to put a piece of round bar in the chuck and put a small piece of flat metal vertically against this. Wind in the tool until it grips the vertical piece against the bar and if the flat piece stays perpendicular you are spot on, if it tips in at the top you are too high and if in at the bottom, you are too low.

CHAPTER 3

CUTTING TOOLS

It was a problem to know which to describe first—methods of holding work, means of holding tools or the tools themselves. Finally I decided to settle for the tools themselves, the reasons being that they are something that you will have to make or purchase at a very early stage and that the type of tools chosen can influence the methods of tool holding. They can easily be accumulated whilst you anxiously await delivery of your lathe and, if you have decided to make them, they can then form little projects to embark on before you take the final step of trying to cut metal.

The tools fall into two types; namely those that will rotate with the lathe and cut material that is held in a stationary position, and those that are held still whilst the work itself is rotated. The latter are the ones about to be discussed and I think it fair to say they are the type of tools normally called lathe tools. What then do we require to cut a chunk of metal revolving in a lathe?. The answer is fairly simple—a piece of metal somewhat harder than that to be cut and with some sort of sharp edge on it. Such an instrument, no matter how crude, will take metal off the workpiece; we can, however make these tools much more efficient if we follow some fundamental rules and if we take just a little care in their construction, when making them at home. To work properly they should

This sketch shows the cutting angles of normal turning tools and should be used in conjunction with the table which gives the correct angle for various materials.

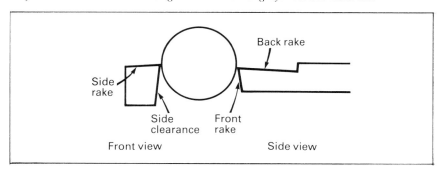

Guide to lathe tool cutting angles

Material	Back Rake	Side Rake	Front Clearance	Side Clearance
Mild Steel	6-10	16	5-9	5-9
	14-22	Zero	5-9	1-3
	15	Zero	5-9	1-2
Silver	6	12	5-9	5-9
Steel	10-12	Zero	5-9	5-9
	5-10	Zero	5-9	1-2
Cast Iron	8	12	5-9	5-9
	6-10	Zero	5-9	1-3
	6	Zero	5-9	1-2
Brass	Zero	Zero	6	5
Gunmetal	Zero	Zero	6	6
	Zero	Zero	6-10	2-3
Copper	8-18	16-25	5-9	5-9
Phosphor	10-20	16-25	5-9	1-3
Bronze	10	Zero	5-9	Zero
Aluminium	8	15-22	5-9	6-10
& similar	8	15-22	5-9	1-2
alloys	10	Zero	5-9	1-2

The first figure given is for roughing out, the second for fine finishing and the third for parting-off operations. In the case of brass there are numerous grades and some of the harder ones will need considerably more clearance.

Chart showing the cutting angles required for various materials likely to be used.

conform to certain basic shapes and, in order that the metal being taken off the job does not clog the tools, there should be relief angles ground into them. This may sound complicated but all it actually means is that if we put our tool against a piece of metal then the edges that are going to do the cutting must recede towards the centre of the tool. These relief angles need not be very steep, and whilst the pundits give recommended ones, these are not too critical and on some materials some of them are not at all important. The tools we shall use will be made of one of three materials. The first of these is high-carbon steel, which is a material that can be hardened and tempered at home (the tempering is to prevent the tool being too brittle and pieces coming off the edges). They can be made of a high-speed steel, which is particularly suitable for what we have in mind. High-speed steel is an alloy and among its constituent metals, apart from steel, it contains tungsten and chromium. I will not go into details of various mixes, but it is sufficient to say that they are all very tough, they grind to a keen long-lasting edge and will stand up to at least 600°(C), without losing their toughness. This is an advantage as the lathe work generates a lot of heat and high-carbon steels, unless handled with considerable care, will discolour and lose their hardness. Finally, the third type are carbide-tipped tools, which are supplied either to fit in holders or as tips brazed onto steel. Creeping into

industry is the use of ceramic-tipped tools, but as far as I can ascertain they are not available for the amateur and the three types refered to above will be the ones we will have to use. Tools ground from steel are commonly known as bits.

CARBON-TIPPED TOOLS

Assuming that we are going to purchase tools then it is not a bad idea to think in terms of at least one carbide-tipped tool. Probably the best for amateur use is a holder that the tip clamps onto. This can be arranged in such a way that the exact height is known and this related to centre height. Enough packing can be kept with the holder so that when mounting it in a tool-post the height will be exactly on centre. All the tips are of the same thickness and so, when assembled, the height of the cutting edge remains constant. If, like me, you are too mean to splash out on such luxuries or, again, if like me, you can't afford it anyway, then the tips are available with holes through them. All that is required is a bar of metal of suitable height with a hole of the same size as the one in the tip. A bolt through the tip into the bar results in an easily assembled cutting tool. The tipped tools with permanently fixed tips are also worth consideration. They come in quite a variety of shapes and sizes and there are bound to be suitable ones for your purpose. They are very long-lasting but do suffer from the disadvantage that they are hard to sharpen. The correct way is to grind them with a green-grit wheel and hone them with a diamond lap. The green-grit wheel is easily available but diamond laps are a different proposition altogether as, apart from anything else, they are extremely expensive. I have found, though, that the light

Left *The author (right) advising on the use of a belt sander to grind lathe tools.*
Below *Carbide-tipped tools. The top one is a holder that takes interchangeable tool bits. Below are two tools with permanently fixed tips.*

A home-made tipped tool holder. It consists of a bar of square steel to which a carbide tip with a hole in it is bolted.

grey stones that are made of silicone carbide will put a reasonable edge on such a tool if used with water. Another dodge that seems to work is to sharpen the tool with a standard, carbide, sanding disc as used in a do-it-yourself drill on a rubber backing disc, as though on an ordinary grindstone. Carbide discs are coloured brown.

CARBON-STEEL TOOLS

Let us consider next the carbon-steel tools which, until a couple of generations ago, were the bread and butter of the turner. Is there really any place for them in today's workshop? I think there is but not when bought in a set as so often happens. Buying sets of tools is generally pointless. It may seem a way to save money or to get a set of neat-looking tools which are all the same. Saving money is doubtful as it is almost certain that one or more of them will never be used. Also, there are many times when tools are required for a particular job and something a little bit out of the ordinary is needed in the way of shapes. Such tools are easily filed to shape from high-carbon or silver steel and when hardened and tempered do a first-class job. Hardening and tempering are quite simple operations. To harden, simply heat the steel until it is bright red and then quench it in oil or water. Clean the metal up and when it is bright again put it on a small tray of sand. Heat the sand until the steel starts to turn a light straw colour and quench again. Hey Presto! a properly hardened tool that is not too brittle, ready for sharpening and use. Silver steel is sold in a whole range of sizes and as both square or round bars. Whilst at first the square bars seem to be the obvious thing to use, it is well worth considering using a round bar of smaller diameter in a mild-steel holder. This again can ease setting-up problems and the holder is available for any tool that you wish to make.

Above right *A sketch showing the shapes of commonly used turning tools as seen when viewed from the top.*

Right *A commercially made tool holder that holds the tool at an angle, thus allowing for easy height adjustment.*

HIGH-SPEED STEEL TOOLS

Probably the most popular material for tools is high-speed steel. It is sold in a variety of sizes both round and square, similar to silver steel. It is already hardened and tempered and heat treatment for this type of steel is not really a viable proposition in the home workshop. It has to be ground to shape but has long-lasting properties and does not complain, keeping its temper nicely! Once again the use of a round bar in a simple holder is to be recommended. It is possible to purchase ready-made tools in high-speed steel; generally, these are marketed by lathe manufacturers and are suitable for any lathes. They are easily sharpened with a grinding wheel or, if such an item is not yours, then with a grinding disc on a rubber pad. Some time ago a belt sander came my way and I now find that it is used in place of my grinder for tool sharpening.

TOOL SHAPES

The standard tool shapes more or less speak for themselves as far as uses are concerned, with knife edges being used to cut sharp shoulders, and round-nosed tools being used for heavy cuts and fine finishes. These uses for the latter tool may seem a little odd; however, in spite of the rigidity of the lathe, tools will suddenly, for no apparent reason, dig into the metal, throwing the metal out of true and possibly even causing it to leave the chuck. This completely ruins the work. A round-nosed tool, because of its shape, tends to do this less as its cutting surface is in contact with a much larger area of the surface of the metal, and so will cut deeper into the metal without catching, than will one with a sharper point. Such catching is very prevalent on parting tools, where only a very narrow section is in contact with the work and the tool is being wound into the metal, rather than along it. As far as finish is concerned, what happens is that as

A standard parting-tool holder which uses a thin strip of high-speed steel bolted to one side of the holder.

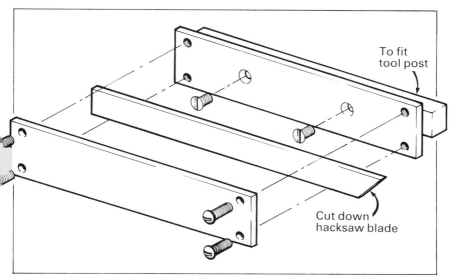

To fit
tool post

Cut down
hacksaw blade

Drawing showing construction of simple parting-tool holder capable of using old hacksaw blades.

the round nose covers a greater area than the feed of the machine, at no time is any part of the metal being turned not in contact with the tool. For example, if the machine feed is a half millimetre per revolution then the tool nose should be in contact with at least three quarters of a millimetre of work. At one time a sprung tool was very popular for finishing but for some reason it has gone out of fashion. Even so, if you happen to come across one they are well worth having.

For a long time I fought shy of 'parting off' in my lathe. Parting off involves putting a very thin tool through a revolving bar of metal. The tool easily breaks if one is not careful and I have memories of an engineering apprenticeship which resounded to the noise of parting-off tools snapping off as they were dug carelessly into the work. I used to take the work out of the machine, saw it off and then face up the sawn end. All very well, except that although I am the proud possessor of two hacksaws, they both insist on cutting across metal at an angle of 45°. This meant I either went undersize and had to start again or that I cut the job so far oversize that facing it was a major turning operation. Finally I took my courage in both hands and, remembering what I had been taught, I sailed forth through a two-inch mild steel bar without any real problems at all. What was the difference between then and now? Two things—first, the lathe was my own, as was the work, so I took a great deal more care and, second, and probably more important, steel has improved. 'Free cutting' now means just that. No more hard spots, just pleasant stuff that, correctly treated, will cut like butter. The correct treatment? A good sharp tool, well adjusted lathe slides, tool tightened well down in the holder and plenty of cutting fluid. The mention of cutting fluid should bring me to the next paragraph but, before starting that, some thoughts about parting tools.

The finished parting-tool holder for old hack-saw blades.

I believe these are best purchased. They can either be bought as a ground tool or as a piece of steel specially shaped to fit in a holder designed to hold it rigid. However, owners of small lathes might find that such tools seem to be unnecessarily clumsy in their machines. For them, a simple parting-tool holder using an old hacksaw blade as the cutting blade is the answer. Cheap and easy to make, I use mine on my large Myford as well as on my small lathe. Remember, though, that you must use high-speed-steel hacksaw blades.

CUTTING FLUIDS

We have now covered all the basic tools for normal turning, but mention should be made here of cutting fluids. These are as much a part of good machining as are the tools. They have two purposes, to cool down the work and to aid the cutting tool. The cooling-down idea is all right in theory but can be difficult at home, particulary in the kitchen or bedroom. A great deal of cutting fluid is required and, as soon as it touches the revolving metal, it does its best to see where else it can get to. The result is often the sight of a lathe operator apparently sitting in the middle of a fountain. In industry, the fluids are processed by circulating pumps; after use, the fluid drains back into a sump and is then recirculated. That is unless you have apprentices about, when the pump nozzle gets turned away from the machine (quite accidentally, of course) and the fluid, under pressure, strikes the apprentice's mate or enemy some ten or twelve paces away. Now you can see why parting tools break so easily!

At home such a system is possible but in most cases the liquid is not recirculated and is simply allowed to go to waste. Cutting fluid can be applied to the work in a variety of ways, either from a tank via a pipe with a tap that will allow it to feed on the work or squeezed on via a liquid-soap container or put on with a brush. The latter I find is an excellent way to get bald brushes, as invariably the hairs get caught in the work and pulled out. Some form of drip feed, then, is probably the best method at present available.

The coolant for steel should be a cutting oil, used in accordance with the manufacturer's instructions. If you find such drip feeds messy, as indeed I do, then there are now excellent oils on the market that need no diluting and which because of their heavy viscosity, are far less messy. There are also some available in spray cans. If you are a car owner then try de-icing fluid, as used for car windscreens, which is excellent. For a good finish, a cutting oil is essential on steel. Brass can usually be machined without any fluid or if it is felt something is needed then paraffin is best. Aluminium alloys are difficult to machine and there is a tendency for metal to build up on the cutting edges of the tool, which means loss of finish or probable breakage of a tool when parting off. It is difficult to stop this, as alloys vary so much, but I have found that one of the de-rusting sprays such as WD 40 can be fairly efficient—failing this, white spirit is quite good. Bronzes can be treated like brass, but cast iron must be machined dry. Any cutting fluid will cause it to harden, with disastrous results. Washing-up liquid gives excellent results with copper.

Before closing on this rather lengthy subject of cutting metal in the lathe a thought should be given to jobs where several items are required to be made the same. If these are small, then a form tool (a tool half the required shape) should be considered. It can be made in two ways—either by turning it from the round, or by filing or grinding it from a flat piece of steel. The tool is taken into the side of the job to a depth determined on the cross-slide dial and all the workpieces cut with its aid should be the same. If you only have four or five items to make from brass it is possible to use a piece of ordinary mild steel as the tool. Whatever type of tool is made, it must have clearance angles or it will be useless.

Form tools are used when making a number of items that all need to be the same shape. The one on the left gives a 3mm radius and ground on the other end is a 2mm radius, which can be used by reversing the tool.

CHAPTER 4

HOLDING THE WORK

Still working with the idea that the work will revolve in the lathe whilst the tool remains stationary, thought must now be given as to how the work will be secured. Work that is not held properly soon makes its escape and departs from the lathe, causing the operator to take evasive action and to try and convince himself that the dents that have appeared where it hit the floor really do enhance the appearance. This usually happens just as the work is being finished off. So whatever we use to hold the work to the lathe, the fixing must be secure.

A three-jaw, lever-scroll chuck as fitted to a Toyo ML1 lathe.

THREE-JAW CHUCKS

The majority of inexperienced people purchasing a lathe for the first time immediately opt for a three-jaw, self-centring chuck of one type or another. This is not really a wise decision, but it is probably prompted by the thought that it is easy to just put the work in the jaws, tighten up and begin. As usual, however, there are snags. First, most three-jaw chucks are only approximate in their accuracy. The amount they are out of true will vary, but out they will be. Even if they start more or less right, they soon wear and what accuracy they once had is lost. Dirt too, of which there is plenty around a lathe, will cause them to lose their accuracy, getting, as it does, into the threads or behind the backplate. This does not mean that the three-jaw, self-centring chuck does not have a place in our plans, because it does. It should, however, only be considered as a second chuck, the four-jaw being the number one. It is possible, in spite of these deficiencies, to be completely accurate with a three-jaw chuck, by using a bar of material of a larger diameter than is required. This is then turned to the desired diameters and any necessary drilling and boring is done without removing the work from the chuck. If the metal is then parted off, complete accuracy must be the result. However, if it is the case that the other end of the metal needs to be worked on, then, once the work is reversed, the accuracy will be completely lost.

A geared-scroll, three-jaw chuck fitted to a Myford lathe. The photograph also shows the use of a steady which will be referred to later. (Photograph by courtesy of Myford Ltd.)

There are various types of three-jaw, self-centring chucks available. As far as the amateur is concerned, the choice will probably be between a lever scroll or geared scroll. The geared scroll will be usual to the larger lathes, while most of the smaller ones have a lever type. The lever scroll is quite adequate for these machines but the strength of grip is not really suitable for work of over about 50mm diameter. All chucks are fitted to the mandrel of the lathe in such a way that they can be removed as required. More often than not, the three-jaw, geared-scroll type must first be fitted to a backplate, particulary if the lathe fitting is itself a screw thread on the mandrel. The chuck will come minus the backplate which is usually obtained as a separate casting, though it is sometimes supplied with the lathe. It is wise to obtain a casting that is finished to suit the fixing of your particular lathe. It is then a question of machining the casting to fit the chuck. How careful you are will decide how well your chuck will serve you in the future. The first thing to do is to machine the face of the backplate to ensure that it is square. The outside diameter then needs to be turned until it is a very good fit to the recess in the chuck. The back of the casting should also be turned across the face. You will now have to mark out the fixing holes, so that the backplate can be drilled for bolts to go through into the chuck itself. Mark out from the outside diameter of the backplate, measuring the distance *in* rather than trying to get a measurement from the centre. To ensure accuracy, check and check again before drilling the holes. Finally before we leave the subject of three-jaw chucks, when buying one, get outside and inside jaws as, sooner or later, you will want to turn something a bit bigger than the largest opening of the inside jaws. Remember also that if you have a large piece of metal with a hole in it, which is to be turned but which will not fit into the chuck, it may be possible to support it by opening the jaws outwards and gripping the hole on the inside. Four-jaw chucks are also reversible.

TURNING BETWEEN CENTRES

One of the most fascinating aspects of turning is working between centres. All lathes should have at least two centres with morse tapers to fit the mandrel and the tailstock, the one that is to be fitted in the tailstock being hardened, as it will have to stand a great deal of wear. Working between centres really means hanging the work between these two centres whilst turning operations are carried out on it. This is done by making small pointed holes in the end of the work into which the centres will go. The work itself is driven by a short bar of metal fitted either to a special plate, called a catchplate, or in the faceplate, which is then made to strike on gadgets called 'driving dogs'. This sounds very complicated but in fact is remarkably simple. First, though, we need to get the small holes centred in the ends of the work. This is done by measuring to the centre of the work, centre-punching it and then drilling it with a centre or Slocombe drill. Finding the centre is the hardest part, but it can be done by very careful measuring from three or four different points and scribing lines right across the work. These lines should all meet at the same point. It can also be done by laying the bar on a pair of 'vee' blocks and setting a scribing block to as near the centre height as possible. Maintaining that height, scribe lines across the end, rotating the bar three or four times. The centre of the bar will become apparent where the lines meet. There are some tools available to assist the process. The old-fashioned bell punch now seems to have gone out of fashion but the combination square can be used quite successfully. When drilling the centre hole remember that this is the means of supporting and driving the bar and they should be of sufficient depth for this support to be given.

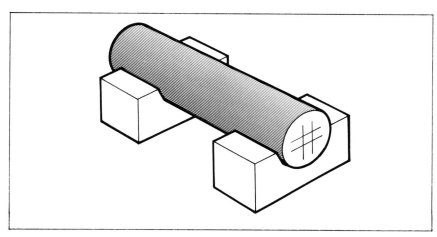

Above *Marks made on the end of a bar for centring by supporting the bar on a vee block and scribing across the end in four places. The centre of the lines scribed is the bar centre. Note that the scribing must be done with a scribing block in order that the height of the scriber remains constant.*

Below *Centre or Slocombe drills as used to make centre holes to support a bar on lathe centres for turning.*

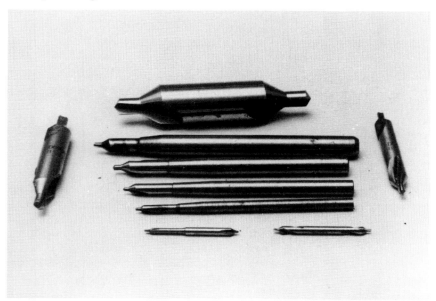

THE LATHE CARRIER

We now need a lathe carrier, as it is called, to drive the work round; this is a simple device for clamping on to the work which is then driven by the bar in the faceplate. Carriers are sold by manufacturers as castings, ready machined. There is no need to go to this expense though. Simply visit your nearest scrap-yard and buy some large nuts. Drill and tap them and then you have your carriers, once a bolt has been put through. Only one tapped hole is required, the bolt both clamping onto the work and being driven. A flat must be filed on the work to give the bolt a grip, otherwise it will just slip round. Before mounting the work between centres for turning, the fixed hard centre in the tailstock should be lubricated or it will overheat and weld itself to the work. Lard used to be the recommended lubricant for this but these days a graphite-based grease is the best bet. It is possible to purchase revolving centres to put in the tailstock and these obviously prevent heating. They do tend, though, to be inaccurate at times and so, if you do buy one, it is as well to get as good a one as can be afforded.

When the work is supported, it will have to be tightened by winding in the tailstock. If it is too loose, it will cause the tool to chatter, and if it is too tight, it will overheat. An old turner I knew used to put a piece of postcard in the hole and then wind in the centre. He could take it up quite tight. The card would wear out immediately leaving the centre nicely adjusted. One obvious disadvantage of turning between centres is the fact that it is not easily possible to face the ends of the work. This can be overcome to a degree by the use of a half centre, or by using a ball bearing in the place of a centre. Simply drill a small centre in a metal bar (preferably bronze or brass) supported in the tailstock chuck, using the Slocombe drill in the headstock chuck. A ball bearing placed between this and the work makes an excellent centre and leaves room to work at the ends.

Lathe carriers. The one in the centre is as purchased, the other two are just old nuts drilled and tapped to take a bolt.

Above *The catchplate from a Myford lathe.*
Below *A faceplate converted to a catchplate by insertion of a bolt and using a home-made carrier to drive the work.*

The bolt, shown from the rear, which converts the faceplate into a catchplate.

THE FACEPLATE

The next piece of equipment to be thought of, with which we can hold our work, is the faceplate. For many years turning between centres and using the faceplate were the only real options that the turner had. With the development of chucks, the use of the faceplate appears to have gone out of fashion, which is a great pity as it is so versatile. Consisting of a disc with a number of holes and slots, it is fairly easy to bolt work to it and it is particularly useful for machining odd-shaped castings. The first thing to do, however, when the lathe is received—and this is extremely important—is to take a light cut across the face of the faceplate. Although this will have been done at the factory, rarely are they true when delivered. This may be because the lathe fitting throws them out a little or because the cast iron tends to warp after a period of time. Whatever the reason, before using it, it is essential that it be trued up.

The most obvious way to fix things to the faceplate would seem, at first sight, to merely bolt whatever is to be turned directly to the faceplate and this is, indeed, what is done. If the casting, or whatever is being turned, has holes in it then bolts can be passed through and if they line up with the holes in the faceplate then you are in business. Personally, I never find that the things line up and so slightly different methods are required. Straightforward clamps work reasonably well but you will need a piece of metal of suitable thickness underneath the clamps to counteract the height of the work. Again not much of a problem, until it comes to truing the work up. My experience is that, as the bolts are slackened off to make the adjustments, out fall the bits of metal. The whole thing ends up like a juggling act. To prevent this it is as well to make up clamps with bolts in the ends that will take the place of the packing, or to make up 'L' shaped brackets using the short leg in the place of packing.

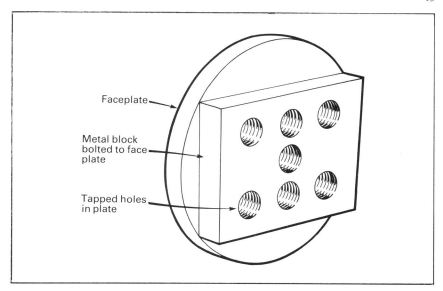

Faceplate

Metal block
bolted to face
plate

Tapped holes
in plate

Above *Drawing showing a metal plate secured to faceplate to hold work.*

Below *A model locomotive cylinder block mounted on the faceplate and supported with an angle plate. Note the packing pieces used to make the pressure of the clamps press down on to the cylinder block.*

The small brackets convert a faceplate into a chuck.

A useful way of supporting work is to make up what amount to little chuck jaws. Square metal pieces are suitably shaped at the end to enable them to seat in the faceplate slots, with a bolt to tighten them up. Another bolt going through the block at right angles clamps onto the work. We therefore have movable jaws with an adjustment, making for easy setting up. Angle-plates bolted to the faceplate make useful supports as can vee blocks. It is really a case of common sense, but do make sure that the work is absolutely secure. If the work is at all off centre then a suitable weight must be placed opposite to counterbalance the vibration likely to be set up and which would otherwise soon wear out the bearings of the lathe.

The converted faceplate in use as a chuck.

Even all the devices recorded above do not always solve the problems of mounting work on the faceplate. A couple of tricks are worth mentioning though. The first is to get a piece of steel plate of a size where the corners come just short of the diameter of the faceplate. Drill and tap it, and bolt it to the faceplate from the back. You can now drill and tap holes for holding bolts just where you like, making work-holding very easy. It does not matter if it gets scored by the tools, it can be used over and over again. Before first using it a skim must be taken across it in the same way as was done with the faceplate, in order to make sure it is true. The other idea is virtually the same, except that the backing piece is made of hardwood and this is used in the same way as the steel plate. Work can also be glued to this and turned, taking light cuts only. Often it is easier to assemble work to the faceplate off the lathe, to be fixed to the mandrel afterwards.

FOUR-JAW CHUCK

The last piece of equipment that I intend to describe at length is the four-jaw independent chuck which is, to my mind, the most useful piece of equipment ever designed. It has far greater gripping power than does a self-centring chuck and, once you have learned to set it up, it is easy to use. The jaws are usually reversible for inside or outside use and, being independent, they can be used for odd shapes in any combination of insides or outsides. Setting up work in it is just a case of checking the tool distance from each of the four jaws and releasing and tightening them as desired. You should not tighten them up too much until the work is set and then when it is right, go round and tighten each one up, a little at a time, until they are nearly right. Test again. The work can then be made to come true by an extra tightening on the correct jaw.

A Toyo four-jaw chuck seen from the rear. Fixing is by locating the chuck in a recess and putting set screws through the holes to bolt it down. Many chucks screw direct to the mandrel.

COLLET CHUCKS

There are many other forms of chuck available but they are not usually the sort of things used by amateurs. One thing that is likely to be encountered is a collet chuck. Collets are highly accurate and very expensive. They are the only answer to the sort of accuracy required by, say, a watchmaker. Each collet takes a set size of material and so a whole set will be required if they are to be used to any extent. They are quite difficult to make and may be beyond the scope of the beginner. The secret is to turn the whole collet and bore into it, in one go, the taper that is required to fit the lathe mandrel. Three slots are then cut in the collet and the end tapped. A suitable bar is made up to draw the collet through into the mandrel by winding the male thread on the draw bar into the female thread in the collet. They should really be hardened and tempered, but if being used on brass only will last for a long time even if just made in mild steel.

A set of collets for a Toyo ML1 lathe.

CHAPTER 5

TOOL-POSTS

Before finally deciding on the type of tools that we will use it is as well to give some thought as to how they will be secured to the lathe. In fact, all we use for securing the tools is a simple clamping device, called a 'tool-post'. It is most unlikely that when you purchase a lathe some sort of tool-post will not be supplied. All new ones will certainly have one, as will most second-hand machines unless you have gone for something really old and decrepit. The type of tool-post supplied varies considerably and we had better look at the various types because it may well be that, even if your lathe has one fitted, you will wish to make a change for something more convenient. Tool-posts also make excellent

A simple, home-made tool-post. One square hole cut out for square tools and one, on the other side, cut for round tools.

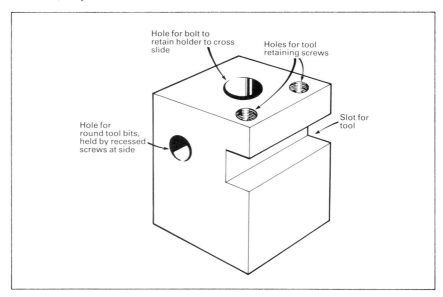

projects for construction, particularly for beginners as, whilst there is obviously need for some care, they do not have to be quite as accurately machined as many other projects. This does not mean that I do not advocate accuracy at all times, it is simply the case that if you are not used to working to close limits then the construction of a tool-post is a good way to start, it being very difficult to mess it up completely.

The tool-post supplied with many lathes consists of a clamp that goes on a central pillar on the top slide, which is tightened up on the tool and holds it very firmly in position. Fine, you might say, what do I need anything else for? Unfortunately this device is highly inconvenient, as it requires either tools set in holders of a suitable size or very large tools. These are expensive items. It also offers no means of adjusting the height of the tools, which have to be packed up with shims. I keep a box full of these pieces, but although there are probably a hundred or more I never seem to have quite the right one. Also it becomes a juggling act when about six or seven pieces are piled on top of each other, with the tool on top. They also have a nasty habit of compressing down, which means that enough packing has to be put in for the tool to be above centre height. The clamp is tightened down and you then check it, to see if you have got it right. If you have, you are so lucky it is hardly worth buying a lathe as you are sure to win the pools and all your time will be spent on world cruises!

A four-way tool-post mounted on a lathe. There are two tools in it and it could hold four if required. Care must be taken to avoid catching oneself on sharp tool edges when more than one tool is in use.

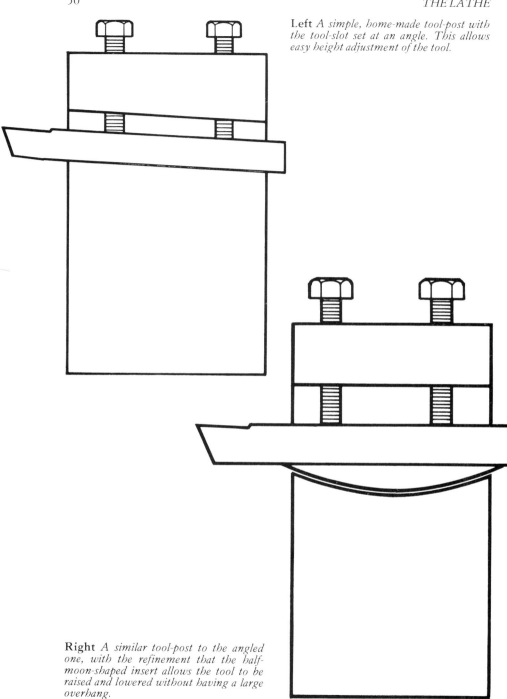

Left *A simple, home-made tool-post with the tool-slot set at an angle. This allows easy height adjustment of the tool.*

Right *A similar tool-post to the angled one, with the refinement that the half-moon-shaped insert allows the tool to be raised and lowered without having a large overhang.*

MAKING YOUR OWN TOOL-POST

What is needed, then, is a means of using less packing and if you do not have a tool-holder, smaller and cheaper tools. All that are needed for such an item is a decent chunk of metal of suitable size, with a hole drilled through its centre (after you have made sure it is square) and with either a slot if square tools are in use or a hole if round ones, and a couple of securing bolts to stop the tool moving; after making this piece of equipment, you are in business, The hole through the middle is best made by clamping the block square on the top-slide and putting the drill in the chuck. The slot or the hole for the tool, as the case may be, is then easily made by mounting the block in an upright position, using the hole that has just been drilled and either running a milling cutter along it or a drill through it. This type of tool-post can be extended by milling four slots, one on each edge, giving a four-tool-post or, if you are likely to use both square and round tools, having a slot on one side and a hole on the other. If a four-tool post is made then, in theory, four tools can be put in it, it can be rotated and the tools used as required. I say 'in theory' as my experience is that when doing this you are likely to cut your hands on the tools facing the rear position. I prefer to use only two tools in the post. It is possible to purchase castings to construct tool-posts should you so desire.

IMPROVED TOOL-POSTS

We now have a tool-post that requires less packing and is far less fiddly to use than its commercial equivalent. But how about one that requires no packing at all. If we take the same basic, single tool-post that I have just described and set the slot at a slight angle pointing up towards the lathe centres, then all we have to do for adjustment is pull the tool out to get it higher and push it in to lower it. Simple really, but as usual there is a snag. The idea works very well except that the tool, if pulled out too far, will suffer from chatter, so the right size tool will be required to suit the post. There is one other type that works on a similar principal, but the same problems do not arise. This consists of a slot with a piece cut away and a half-moon-shaped packing piece. Rotating the packing piece raises or lowers the tool as required. It is difficult to visualise this arrangement from a description, but it should be clear from the drawing.

A useful piece of equipment is a rear tool-post. This, as its name implies, goes at the back of the lathe. It will not fit on the top-slide but should be bolted to the cross-slide. The tool goes upside-down. In construction, it is similar to the simple, single tool-post, and is particularly useful for parting off. The forces acting on the lathe drive downwards which gives a greater rigidity and enables a heavier cut to be taken, whether turning or parting. It is easier to use a rear tool-post with an extra long cross-slide, but it still has its uses with a standard cross-slide.

There are various patent tool-posts on the market that are advertised from time to time. Most of these work on a quick-change principle or have various means of adjustment. The only way to decide whether these are to your liking is to have a look at them and ask yourself whether or not there will be any great advantage in getting one. Much will depend on the type of work you are going to use it for, apart from which, no matter what you buy, you will always be proud of the one you make.

Particularly useful for parting-off—a rear tool-post. Similar to the normal type of tool-post it is mounted at the back of the cross-slide and pressure applied to the work is in a downward direction which gives greater stability.

CHAPTER 6

TURNING, DRILLING AND BORING

Readers who have struggled this far will no doubt be very pleased to find out that we are now going to get on with how to do it, rather than what to do it with.

THE BEST WAY TO TURN

Assuming we have set up our round stock in the chuck or collet, the time has come to start turning bits off it. It may well be that it is part of some larger operation or it could be just for the pure satisfaction of seeing metal being cut. The latter was a very popular operation with students when I was teaching. Give them a bar of metal and leave them to it and they would turn away for hours. What they most liked to see was the swarf coming off in as long a piece as possible. Whilst in the students' case this was due to idle curiosity, it was in fact good basic work. Particularly when turning steel, the longer the swarf the better, as it shows that things are going well. What do we need to achieve this? First, a well sharpened tool of more or less correct profile, second, the tool firmly tightened in the tool-post, with as little overhang as possible and, third, an adequate amount of cutting fluid. If there is any great overhang, the metal should be supported by a centre in the tailstock and, if it is a particularly long piece, it should be supported by a 'steady'.

The steady has only been mentioned in passing before. It consists of a gadget that supports the metal and stops it whipping in the middle. After all, considerable force is being exerted by the tool and it is often easier for the metal to move out of the way than it is for the tool to penetrate. Steadies are sold by manufacturers of lathes. They are either of the type that travels along with the carriage or the fixed type that bolts to the bed. They are not too difficult to make. A piece of sheet metal cut to shape provides the main body. A piece of angle-iron rivetted or welded to it makes the bracket to support it on the lathe. The steadying arms need only be some form of nut, either welded on or screwed through from the back, with threaded rod providing the supports. This will not be as good as an expensive commerical steady but it is still quite useful and will provide sufficient support. If you have a one-off job, then drill a hole in a piece of wood and fix it to the saddle. The metal goes through the hole being supported at the rear to prevent it moving away from the tool and, if for some reason this

A home-made steady, made as suggested in the text from sheet steel and angle iron with, in this case, nylon adjustable arms.

Below right A piece of wood being used as a steady. The wood is bolted to the saddle with a piece of angle iron. It can be used in reverse to support a drill, and helps prevent this wandering by passing the drill through it between the tailstock and the work.

cannot be used, then a piece of wood just rubbing along the back of the metal will support it. Needless to say, a steady must be adjusted to suit the metal after a cut has been taken.

Having satisfied all these requirements, we can at last start cutting. Select a suitable speed for your lathe—the details of cutting speeds should be referred to here. Unless you have a very expensive machine there will probably have to be a compromise and the nearest lower speed should be selected. If you do not know the lathe speeds then work on the principle of the larger the diameter the slower the speed and the harder the material the slower the speed. A piece of 50mm diameter brass, then, turns at a faster speed than the same diameter steel. A smaller diameter piece of steel will work at the same speed as the larger diameter brass. The system sounds very complicated but with a little experimenting you will soon understand it. Most full-time turners do most of their work by feel and never refer to charts or calculations. Now apply cutting fluid if it is to be used and wind in the tool; take a light cut first of all and then make the next one deeper. Take the diameter down to just slightly larger than that required, then either resharpen the tool in use or use a new one to finish off. If cutting to a sharp corner use a round tool first and then a knife edge to take out the corner. A little at a time is the rule in doing this, as both the length and depth will need to be brought to the correct size. The use of files and emery cloth to finish off is not a good practice and it should be possible to get a perfect finish with a well-sharpened tool.

If the result is not smooth then there is probably some looseness somewhere. Either the tool or the work is vibrating. This will possibly be indicated by a screaming noise when the work is being turned. It is also advisable to check the tool height because, if packing has been used, the tool could have gone lower than thought. One problem is how to reset the tool when it has been

This chart can be used to calculate approximate cutting speeds. The column on the left represents the diameter in millimetres of either the material revolving in the chuck or the cutting tool which is rotating. The centre column gives the speed of rotation of the lathe, while the one on the right gives the speed of rotation of the material. Draw a line from the material column to the diameter column; the cutting speed required is the nearest one to the point of intersection in the centre column. For example, a line from 'brass' to '25mm' (the diameter of the material) will pass through '900', which is the speed at which the lathe should be rotating. If a 10mm diameter milling cutter is in use on mild steel the line will pass through '700', which again is the required speed in revolutions per minute. If the line passes half way between cutting speeds use the lower one unless your lathe has a suitable speed between the two.

mm	Cutting speed	Material
		aluminium
5	3,000	brass
		copper
	2,000	
10	1,000	
15	900	gunmetal
20	700	stainless steel
	500	
25	400	
	300	phosphor bronze
	200	mild steel
50		
	100	
75	80	
	60	
100	50	cast iron
	40	silver steel
125	30	
150		cast steel

An unusual type of steady. The metal to be turned down is 2mm-diameter brass. Worked on without support, it would bend. A hole has been drilled in a piece of metal in which the work just fits. The metal can then be either filed or milled away leaving the work exposed for the turning operation but unable to move because it is supported on three sides. If such a steady is to be used then the work must be well lubricated to prevent seizing up. If the possibility of damaging the work by marking it exists, then use a piece of plastic to make the steady but use a large diameter so that the steady itself does not move out of line. The other end of the steady is supported in a chuck in the tailstock.

resharpened to prevent it from digging in too deep. A piece of ordinary wet cigarette-paper round the work will do the trick. Once the tool catches the paper then you can say that it is just about to touch the metal. When winding the tool back along the carriage take it away from the work first. Use the dials to tell you how far you need to wind back. Remember it is proposed to work to very fine limits and if the tool is wound back whilst in contact with the work, with the lathe running, it could mean a loss of accuracy. The reason for this is that the screw used for winding the tool along wears on its front face. When wound back it meets a different surface and this is inclined to push it inwards. If there are no dials on the machine and you need to keep the tool in position for accuracy, then turn off the motor and wind the tool back with the work stationary. A very slight score mark may be the result but it is the lesser of two evils.

The basic sequence of turning operations should be to turn the larger diameters first and follow with the small ones, and then to carry out any boring and drilling that may be required. Profile work, such as rounded shoulders on external surfaces, should be left until last if possible. All the suggestions given above for turning are generally applicable to the boring operations shortly to be described

INS AND OUTS OF DRILLING

If we start by thinking in terms of drilling small holes then a tailstock chuck is almost a necessity. The old timers did not have such luxuries and it is possible to manage without one by holding the drill bit in some sort of clamp and using the tailstock centre to supply the force. This is not a very satisfactory method and the use of a tailstock chuck makes life easier. All holes should be started by using a centre or Slocombe drill, which come in a variety of sizes. Ideally, one with a pilot slightly smaller than the drill to be used should be selected and this should be taken into the work far enough for the main body to leave a slight countersink, which will now provide a nice starting point for the drill.

These centre drills are remarkable tools. Even if in some way you have managed to start slightly off centre the tool will usually straighten itself up and become quite accurate. The drill can be wound in next. Take it easy, as broken drills are almost impossible to extract and when you have spent several hours machining a piece of metal only to break the drill in the final hole, you might be inclined to say 'bother'. Keep withdrawing the drill in order to clear the swarf. Drilling should be done in stages using progressively larger drills each time. Good quantities of cutting lubricant help quite considerably and, on steel, make sure the drills are sharp. In the case of brass and copper there is considerable danger of the drill snatching if it is too sharp and it is wise to just rub a piece of emery cloth over the cutting edges—this should ensure that snatching will not take place.

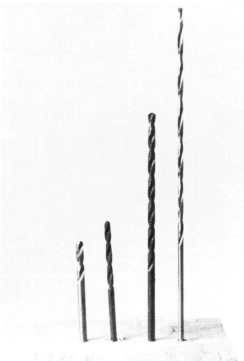

Drills come in a variety of lengths. **Left to right:** *stub, standard, long series and extra-long series. If the longer ones are to be used,* **drilled with one of the shorter ones to its full depth,** *as the longer the drill the more it is inclined to wander.*

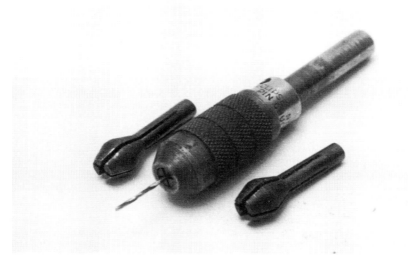

Very small drills need to be put in a pin chuck and then into the tailstock chuck, for greater accuracy. This set of pin chucks with three collets, made by Eclipse, is suitable.

Cutting speeds are as important in drilling as they are in turning, but do not foget that it is the area being cut that decides the speed and in this case the area is the diameter of the drill. The work will, then, need to travel faster than when turning operations are being carried out. It is doubtful if the correct speed will be available, as for small drills it is quite high—also it is not advisable to have an enormous piece of metal in the chuck or on the faceplate spinning round at twice the speed of sound, so some compromise will have to be made. If a lot of work with small holes is to be carried out, then an attachment should be made up to set the drill on the tool-post. This used to be quite a problem and involved major work to construct. It is now possible to purchase excellent little, low-voltage hand-drills. One of these in a suitable holder will do the job quite nicely. The holder does not need to be elaborate and as long as it is absolutely secure can even be made of wood. Very small drills of a fraction of a millimetre, or those in the number 65 to 80 range, should be held in a pin chuck and then put in the tailstock or the tool-post driller. The pin chucks are available with sets of collets to take a range of very tiny drills and they are not as expensive as might be expected. If a hole is to be accurate then it should be drilled just a fraction under size and finished with a reamer. Reamers are expensive and a correctly sized piece of silver steel made into a 'D' bit will do the job just as well. This simple little tool is just a flat filed on the steel to exactly half the diameter with a cutting edge put on the front. The round sides supply suitable cutting edges along the length. The 'D' bit is also the ideal tool for finishing blind holes square at the bottom. It is quite possible to make your own drills—watchmakers have done so for centuries. If used with care, these little drills are quite satisfactory, particularly on soft metals.

Sometimes very long holes are required and drills are sold in various lengths

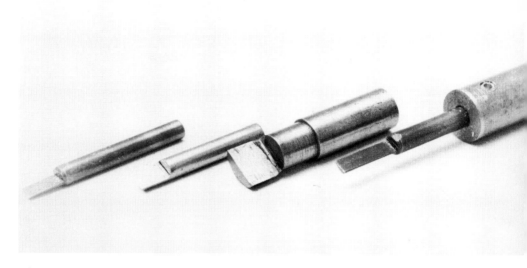

Above *A selection of D bits made for various purposes. They are made from round silver steel and are filed to half the diameter with a cutting edge at the front. The edge should be set at the same angle as the front rake of a normal turning tool.*

Below *A home-made drill, which is a D bit with a point and suitable cutting edges. It can be used at much slower speeds than a normal drill and is less inclined to wander.*

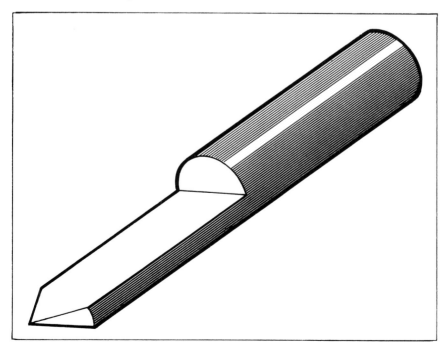

Larger drills have taper shanks to enable them to be put straight in the tailstock. This gives greater support than a chuck and causes less overhang.

for this purpose. The long-series drills do have a nasty habit of wandering off centre, however, and it is advisable to use some sort of steady arrangement on the drill bit and first to use a standard drill, to get the hole started. A simple piece of hardwood with the correct size hole in it and mounted in the tool-post is a good steady, as any sign of wander on the drill will show up and steps can be taken to correct it immediately. Drilling deeper will not result in accuracy getting better, just steadily worsening. Once again the use of different size drills helps but there is probably a lot to be said here for the use of a 'D' bit as it is far more rigid than a drill.

Larger holes can also be made by using drills. The larger sizes are usually made with a morse taper to fit in the tailstock direct. Often these can be purchased cheaply as surplus stock, particularly in Imperial sizes. If you happen to acquire some with a smaller taper than that on the lathe then a sleeve is obtainable to convert the lathe taper to a smaller size. It is not advisable to use an adapter that enlarges the taper as there is a great deal of overhang to cope with in this case, resulting in considerable loss of accuracy.

EXPANDING THE BORE

Very large holes will have to be bored out using the methods described for turning operations, but working from the inside and gradually winding the tool outwards to increase the diameter of the hole. Boring tools consist of long bars with a cutting edge and the photographs will show better than any description how they can be made. The thinner the bar, the more it will tend to whip and make the hole narrower at the end nearest the headstock. The rule is to use as thick a boring bar as possible in order to achieve accuracy. Measuring the diameter of the hole is a problem. Most vernier calipers have a means of measuring inside diameters but this will only give the size at the end. An inside micrometer used in several places will give good results. If such a tool is not available then a bar of metal of exact size will do the job. Remember to turn the test bar to exact size before boring the hole, though, otherwise you will have to take the work out of the chuck, after which you will definitely be unable to put it back in again accurately. Almost certainly, on first measuring, the hole will have a taper and several runs with the tool will be required before it will be true.

One method of getting absolute accuracy with a bored hole is to mount the work on the saddle and bore with a bar between centres. This is just a bar of metal

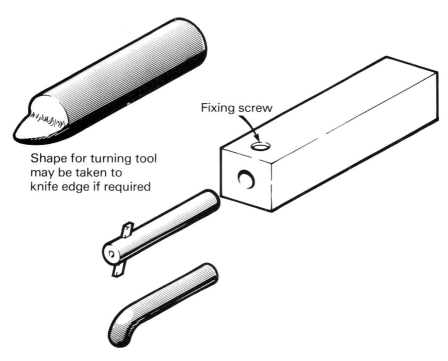

Fixing screw

Shape for turning tool
may be taken to
knife edge if required

Above *A drawing of a simple boring-tool holder consisting of nothing more than a piece of square bar with a hole drilled in it for the boring bar at centre height. A small hole is drilled and tapped crossways for a screw to secure the boring bar. The holder can also be used to hold a normal turning tool.*

Below *A selection of boring tools.*

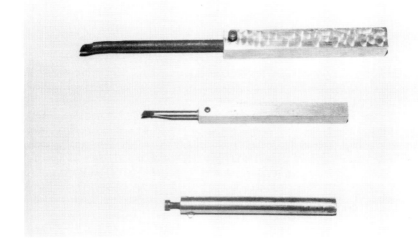

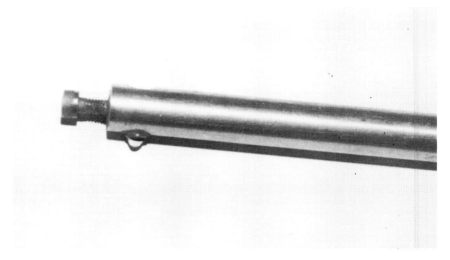

Above *A boring tool consisting of a round bar for setting in a holder and a small tool bit protruding from it. The tool bit is secured with a screw. If blind holes are to be bored and this type of holder has to be used, the tool must come at an angle above and beyond the bar and the screw must be recessed.*

Below *This drawing shows the construction of a between-centres boring bar.*

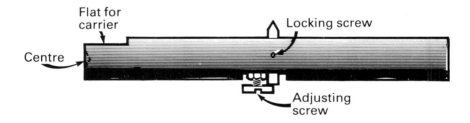

with a centre in each end driven as described for turning between centres. A hole is drilled through the bar and a cutting bit put through and held with a grub screw. If some sort of adjuster can be made, so much the better. The bar is now rotated and the work moved up and down on the saddle. Accuracy is thus assured, and this method should always be used where possible in preference to the ordinary boring bar. Another advantage of the bar between centres is its ability to bore two individual components which should match, such as a pair of cylinders, and to ensure that both bores will be identical.

In order to establish the finish on the bore it will mean looking inside it. A small mirror like that used by a dentist is usually suitable, but if this proves too large or you do not have one, then a piece of broken hand-mirror stuck on a bit of wood is a good substitute. Getting enough light is a problem, but a hand-

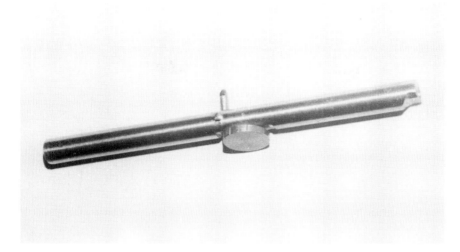

Above *The finished boring bar.*

Below *The between-centres boring bar being used on a small lathe to bore a cylinder block for a model locomotive. The block would be far too large to accommodate on the faceplate of such a small machine, also there is no possibility of the boring bar flexing and making a taper bore. Because it is supported at both ends the bore will finish true.*

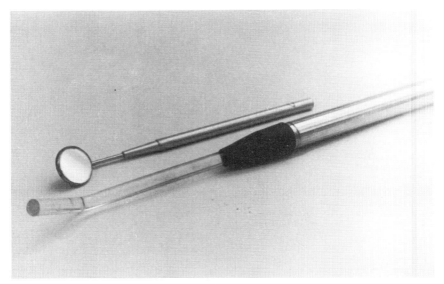

Above *Examination of work that has been bored is difficult. Here we see two useful aids: a torch with a perspex rod that takes light right into the bore and a dental mirror which helps examination.*

Below *Measurement of the bore is a problem. A piece of metal of the exact size will do, but if measurements must be taken for any reason then either a vernier gauge as at the top or an inside micrometer should be used. The vernier suffers from the disadvantage that the spikes at the top are the part used and these will only measure the lip of the bore and will not tell whether it is of true diameter right through.*

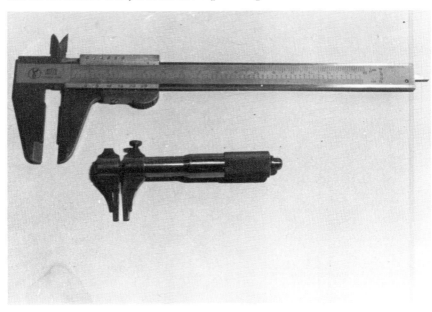

torch of the small, pocket type is useful. It can be made even more useful if a piece of polished, perspex rod can be fixed to it as the light will then travel down the rod and illuminate the inside of the bore. A similar effect can be obtained with a torch and a piece of glass tube of the type used for water gauges. This can be easily and cheaply bought at chemists and is held so as to just touch the torch lens.

Basically, everything that has been written about turning operations also applies to boring. Tools to be used are of the same shape as far as cutting edges are concerned and the only difference between the two operations is that it is better to take lighter cuts when boring. There is no problem when holes go right through work but if the hole is blind it requires a great deal of care and patience to ensure that the tool does not travel too far and hit the end. The fact that, of necessity, boring tools are thinner and longer than ordinary turning tools means that there is a tendency for them to move a little. It is then, advisable to take several cuts on the same cross-slide setting shortly before reaching the ultimate diameter of the bore. This will straighten out any taper caused by tool movement. Final cuts should be very fine indeed and, again, may have to be repeated several times in order to get the bore absolutely true.

It may sometimes be that, with a fairly small bore hole, the bottom of the tool will rub on the work due to the radius of the bore being smaller than the depth of the tool. In such cases a small step will have to be ground on the tool just under the cutting edge to allow for clearance. Do not overdo this, as the less metal on the tool, the weaker it is. If a blind bore has to be left square at the end then stop just short of the full length and finish by taking light cuts across the end starting from the middle and working outwards. This leaves a nice smooth finish. A very small tool at the final stages will ensure a sharp corner at the bottom of the bore. Cutting fluids should always be used where possible and swarf should be cleared from the bore at frequent intervals.

CHAPTER 7

THREADING

It is inevitable that at some time or other it will be necessary to put a thread on the work in the lathe, either internally or externally. We will start with internal threads in the smaller sizes, which more often than not are accomplished by tapping. Threading is a simple operation involving drilling a hole of the correct size (this will be found by reference to charts) and then running in taps. Taps come in three grades, taper, second and plug. This is the order in which they should be used but, if expense is to be considered, then there is no reason why the second tap should not be dispensed with.

Using a tap in the tailstock chuck. This operation requires a great deal of care and patience.

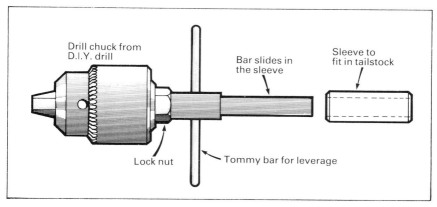

This drawing shows how to make a simple tapping guide for use in the tailstock. It consists of a sleeve, which is held in the tailstock chuck and another chuck, on a bar of metal, which slides in and out. A small tommy bar enables the work to be kept still and the tap rotated if required.

THE ART OF TAPPING

It all sounds very simple but as usual there are some snags. Taps have a nasty habit of doing one of two things. First they seem to take a delight in going into the hole at an angle. This would seem to be an absolute impossibility, but it is surprising how easily it happens. Second, they insist on breaking, usually, like drills, waiting first until all the rest of the job is done. Both these nasty habits are allied to each other and a tap that is put in straight is less likely to break than one that is not. In order to get the tap to enter the drilled hole correctly it needs to be lined up accurately. We have a ready means at our disposal for this purpose in the form of the tailstock chuck and the tap can be held in that. The lathe should be wound round by hand by pulling either on the drive belt, or on the chuck, provided it is secure. Take care, though, if the chuck is of the type that screws direct to the mandrel or it will come off whilst you are in the middle of the job. The taps should only be taken in a very little at a time. Whilst this might seem very time-consuming, it could save a great deal of heartache later on. The tap should be wound fully out and any metal removed from it before taking it back in. Half a turn a time is usually enough. It sometimes happens that it will not be possible to wind the tap out, if it catches. It probably, at the same time, will not go in very far either. This situation calls for a great deal of patience, it being necessary to very gently rock the work backwards and forwards until it clears itself. When a tap has gone into the full depth and is free on the first few turns coming out, it can, unless it is very small, be brought out under power. Never try to take it in under power. Whatever tapping is being done, the use of a cutting compound specially designed for tapping operations is essential. All good tool stockists should be able to supply these compounds.

Sometimes it is not possible to use the tap in the way described and this is likely to be the case where large taps are employed, as there is insufficient torque on the tailstock chuck. Great leverage will be required and a tap wrench will have to be resorted to for extra leverage. The lathe can still be used to ensure that the tap goes accurately into the hole, by putting the tailstock centre in the hole that is almost invariably be found in the end of the larger taps. The assistance of somebody else will be required unless you happen to have three hands or are good with your feet, but the centre, if kept in contact with the tap, will help to keep things square. The ideal tool for tapping and using the tailstock is a bar of metal, either with a taper to fit or go in the chuck and another piece of metal bored out to slide on it. Either a collet, a chuck or a hole with a fixing screw holds the tap and the fact that the tap in its holder is free to move gives it greater flexibility and makes breakages less likely. It is certainly worth using the lathe as a tapping guide in the ways referred to, a far greater deal of accuracy being assured than when the work is taken to the bench and the holes tapped using a tap wrench and no other support.

EXTERNAL THREADING

External threads are put on with dies, it being a case of running a die along the metal which has been turned to the outside diameter of the thread. Again caution is advised and it is as well to wind the die out several times during the operation in order to clear out scraps of metal. A tapping compound should be used. The majority of dies are of the split type and these allow a certain amount of adjustment. By loosening the centre screw in the die-holder and tightening the outer ones, a die can be made to cut a slightly smaller diameter, which gives an easier-running thread. The amount of adjustment to be put on depends largely on the use for which the thread is intended.

The dies should always be set in a die-holder in order that they go on square, as they have similar nasty habits to the taps, with regard to wandering. Die-holders consist of a piece of metal bored out to take the die, sliding on a bar of metal, with a morse taper to go into the tailstock. If you are making your own and do not fancy taper turning, a parallel bar to fit the chuck will do. The piece holding the die should either be knurled or should be drilled and tapped to take a metal bar, which can be used to prevent the holder turning when in use. The illustrations show how this should be done. A holder of this type is a simple

A set of die-holders to take various size dies. Their construction is reasonably simple; they consist of a metal bar turned to fit the taper in the tailstock, with the other end left parallel. The holders themselves are bored to fit the parallel bar and the other end is bored to take the die. A slot in the actual holder locates on a pin in the parallel bar to prevent the die-holder from turning round.

Above *A close-up view of one holder. The tommy bar allows for extra leverage which will be required in the case of large diameter dies.*

Below *A tailstock die-holder in use in a lathe. On this one, a knurled holder provides the grip, making a tommy bar unnecessary. Such an arrangement is only suitable for small threads, where less effort is required to thread the material.*

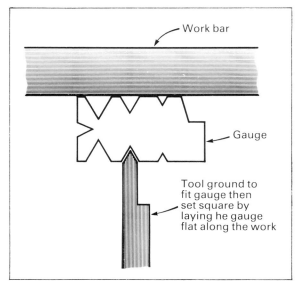

Work bar

Gauge

Tool ground to
fit gauge then
set square by
laying he gauge
flat along the work

Left *A thread gauge for grinding and setting a tool to be used for screw cutting. The tool is ground to the required angle and then set against the work, using the gauge as a guide. The gauge has a number of different cut-outs to suit a variety of threads.*

Below *A thread gauge being used to set a tool in the lathe.*

exercise to construct and although there are no dimensions given, it can easily be made from the details supplied.

If suitable taps and dies are not available, and assuming that the lathe has the facility to cut threads, then screw cutting will have to be resorted to. With large threads, it is as well to screw cut before using a die anyway as too much leverage is required for either a tap or a die to be run straight on to the work. The tool used should be ground to the profile and angle of the thread to be cut. Different threads have different angles and so a chart will have to be referred to. Charts giving thread angles and tapping sizes are very cheap and are essential once work is started. The correct gear ratio will have to be used, either by changing the gears at the end of the lathe or by means of the automatic gear box, if fitted. The tool is set square by means of a gauge, which can also be used as a guide for grinding. This can be purchased, although making one is a fairly simple job. Cutting should be done in easy stages and it is a very good idea to make a groove at the end of the thread first, just slightly below the inside-thread diameter. This is called an undercut and allows the tool to be taken out at the end of the cut without danger of it being caught up. Selection of where to engage the carriage is made by reference to a gadget called a thread dial indicator.

A thread dial indicator enables the tool to be reset at a given point. The tool is withdrawn from the work and, providing the self-acting lever is operated at the correct mark on the indicator, the tool will locate at the correct point of the thread for the next cut.

FINISHING THREADS

The thread should be finished with either a die or a tool called a chaser. Chasers are rarely used these days by model engineers but are still available. If a die is not available then gently clean up the thread with a needle file of suitable shape, remembering that gentle is the way, as heavy filing would spoil the thread. Another way to finish the threads is to apply some Brasso, or a similar substance, on the male and female parts, and screw them backwards and forwards for a while. This will leave a nice finish and a smooth-running thread. Internal screw cutting follows the same principles. The tool is first set square with the gauge on the outside of the work. There must be an undercut with internal threads or disaster will result. If, for any reason, it is not possible for the undercut to be incorporated, then it is possible to cut the thread with the lathe running in reverse direction of rotation—do not try and use it both ways.

If the thread is to be used for driving purposes, such as on a vertical slide, then it should be either square or Acme form, preferably the latter, as vee threads are unsuitable for driving purposes. It may be that a two-start, or any other form of multiple-start thread, will be required. This is a case of putting threads in between each other and the mandrel has to be turned the exact distance round before starting the second thread. This is tricky, and probably the only way to succeed in being accurate is by using a change wheel to make the division, in the method described later for dividing. This project cannot, though, be recommended for complete beginners.

Cutting a two-inch British Standard Pipe Thread on an alloy bar. Note the soft packing in the chuck jaws to prevent marking the work and the use of a revolving centre to support it.

Threads take different forms and these should be checked with the gauge. Most threads for securing items are of the vee form at the top; there is, however, a rounding of the vees at both top and bottom. Square threads and Acme are used for driving appliances such as jacks. The Acme is the most suitable for the sort of work done in the home workshop. It is also the easiest to make.

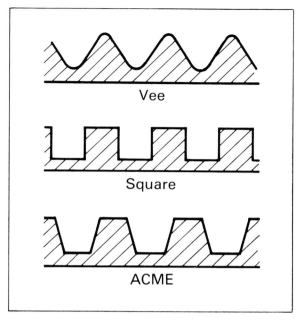

Vee

Square

ACME

CHAPTER 8

MILLING

More and more amateurs are now finding that they can obtain small milling
machines and so the use of the lathe for this purpose is declining. Even so there
are still many people who have to use their lathes for both purposes and even if a
milling machine is available then there are some operations that are better
carried out on the lathe. The difference between milling and turning is that in
the case of the latter the work revolves whilst the tool remains stationary, whilst
in milling operations the tool is made to rotate whilst the work is held steady.
Probably the only exception to this distinction is boring between centres, as
described in the previous chapter. Because the tool is to rotate we usually need to
have tools of a different basic design. I use the word 'usually' because this is not
always true and in fact single-point cutting tools can be made to do milling jobs
quite well, a typical example of this being the fly cutter. This consists of a tool
identical to a lathe tool (usually the round-nosed type) set in a holder that makes
it rotate off centre. A wide cutting arc is made and, when this is slowly traversed
over a large flat surface, a suitable finish will be imparted. Fly cutters are very
easy to make and no home workshop should be without one. If extra rigidity is
required, the tool can be bolted direct to a faceplate and used in that way.

TYPES OF MILLING CUTTERS
Most milling cutters are of a multi-tooth design and they come in two basic
types, the round types with a hole in the centre and the teeth round the edges, or
the long thin types which look somewhat like a drill bit without a point. The first
type, initially, would seem to have no real part in operations on the lathe and for
the owner of very small lathes this is quite true. However with the larger
machines one or two of these can be useful acquisitions, particularly as they can
often be bought very cheaply on the surplus market. They will need to be used
on a mandrel and so, if you do purchase them, it will be as well to make sure
that if more than one is bought they all have the same size hole in the centre. It
is extremely irritating to have to waste a great deal of time and material making
up mandrels for each tool.

The second type is divided into two categories, end mills and slot cutters. The
end mill will probably have at least four cutting edges and of course four flutes,
whilst the slot drill just has two. As only very light cuts can be made on a lathe

Above *Milling cutters.* **Left to right:** *a side and face cutter. This type of cutter has limited uses in the lathe but can be used for making large flat areas on metal; the end-mill and the slot-drill, the uses of which are described in the text; a slot-cutter which is useful for cutting tee slots of the type used on the lathe top slide. An end mill or slot-drill is first used to get the full depth of the slot. The cutter then makes the tee. Only very light cuts can be taken with this type of cutter. The last type is a dovetail cutter which is used, as the name implies, for making dovetails. In front are two slot-drills ground to cut particular profiles. The one on the left cuts a rounded slot while the one on the right has a sixty-degree angle.*

Below *A slitting-saw being used to cut a brass bar. It can be used in any situation where narrow grooves are required.*

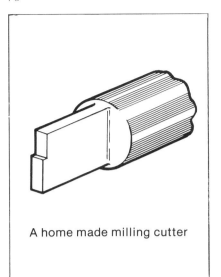

A home made milling cutter

Left *The drawing shows a home-made end-mill. Once again it is similar to a D bit but with both sides removed and with two cutting edges.*

Below *The fly-cutter consists of a small tool similar to a turning tool set in a steel block and rotated in the chuck. The tool can be shaped as a form tool should a particular profile be needed.*

either will do the same job. Out of the two, the slot drill has the advantage for the man working at home who does not have a proper tool-grinding machine. With care, it is possible to sharpen the slot cutter on the grindstone, which you cannot do to an end mill.

One other useful type of cutter that should be mentioned at this stage is the slitting saw. This is a narrow circular saw blade which has to be mounted on a mandrel and can be used for parting off heavy metal, as well as slotting screws and similar jobs. Remember, though, any cutter on a mandrel must be supported by a centre, if it is at any distance from the chuck.

End mills that will be quite suitable for work in the home workshop are easily made. They are made in the same way as our old friend the 'D' bit, except that metal is removed from, and a cutting edge put on, both sides. Any odd shapes that need to be milled out can be made with a home-made form cutter held in a round bar in the chuck. I write chuck but, where possible, it is better to use collets for holding milling cutters.

PREPARING TO MILL

Work to be milled is bolted to the cross-slide in a similar way to which work is bolted to the faceplate. Here an angle-plate is invaluable, as not only does it support the back but it is often possible to bolt work directly to it. One snag with milling in this way is the fact that there is no provision for raising the work in order to take a second cut. For this purpose a vertical slide is used, which will be discussed shortly. It is, however, possible to raise work fastened to an angle-plate by means of tiny jacks and in fact doing it this way has the advantage over the vertical slide, in that the work is held much more rigidly. It follows that the use of jacks will also allow work to be lowered when cutting keyways and similar such operations. The work must be bolted up tight again, after each movement of the jacks.

In order to get full flexibility into milling in the lathe, a vertical slide will be required. Most manufacturers make one to fit their machines. This is not a very complicated device, but is simply designed to allow the work to be raised and lowered vertically. As the top-slide will not be used at the same time, it can be used as a vertical slide, provided some way can be found of bolting it to a sturdy angle-plate which allows it to operate. This is not a particularly difficult expedient to arrange and very successful. Some vertical slides have a device to allow them to be set at an angle (these are called swivelling slides). This certainly is handy, but such designs always lose a certain degree of rigidity. Another problem is that the slides are rather narrow so, if long work is to be mounted on one, some form of metal plate of a reasonable thickness should be bolted behind it, to prevent movement at the edges. Once more our old friend, the angle-plate, is useful here or, failing this, a long piece of angle-iron to mount the work on to stop it moving. The vertical slide must of course be set square to the mandrel, and the best way to do this is to mount the faceplate on the mandrel and wind on the vertical slide with the bolts loose. Adjust it until it lies square against the faceplate. Only light cuts should be taken when using the slide.

It is not to difficult to make a vertical slide and although such a tool may not have the refinements of a commercial product, it is possible to make it of greater width, thus saving the need for packing to strengthen it. If a machine-vice is to

A swivelling vertical slide. (Photograph courtesy Myford Ltd.)

Above *A bar of metal is held in the tool post and a slot milled with an end mill held in the lathe chuck. Note compensating packing piece to ensure that the work is held absolutely firm.*

Below *A cross-slide bolted to a steel block in use as a vertical slide.*

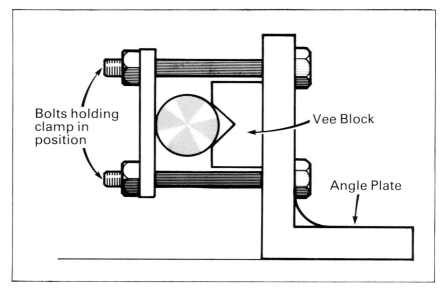

The drawing shows a method of setting a bar on a cross-slide to mill the end. A vee block and angle plate are used.

be used on it there is no need for 'T' slots, as tapped holes where the mountings are required will do as well. They will also serve to bolt work directly to the home-made slide, provided suitable clamps have been made up. I have pointed out the limitations of vertical slides, but this does not mean that the use of one is not worth consideration. Many very fine model engineers never use anything else and a great deal of high-quality work is regularly turned out by their users.

MILLING ATTACHMENTS

In recent years, manufacturers have again tended to produce milling attachments for the lathe. Such gadgets were first made commercially many years ago and the large majority of model engineers also made such an attachment. They consisted simply of a spindle that could be driven by a motor, fixed in such a position that the cutter could be brought to the work on the cross-slide. Often they were driven by a small overhead shaft from the lathe itself. There is no reason why such an attachment should not still be made, but the modern enthusiast has the advantage, as it is easy now to get high-quality, low-voltage motors to drive the attachment. Mains voltages should not be used because of the danger of electric shocks. Having said that, I must just mention one enthusiast I know who quite successfully uses his do-it-yourself electric drill for the purpose, which being of the double-insulated type presents no danger.

Milling attachments supplied by dealers fall into two categories. The first sort are driven by a spindle and fit in the mandrel, while the others have their own independent motors. Both use the cross-slide as the means of traversing the work, and both virtually turn the lathe into a proper milling machine. Both types are a considerable advancement on the vertical slide.

CHAPTER 9

OTHER OPERATIONS

We have mostly covered basic lathe operating as far as is possible within the limited space of a book of this type. There are, though, some situations which have not yet been discussed and I propose briefly to run through these. First, we should think about turning tapers, an operation that will be required if we are to make tools to fit either the headstock or tailstock.

TURNING TAPERS

The actual turning operation differs only in the fact that cuts will need to be light, but the setting up is a problem. The top-slide can usually be set over for short tapers and the tool wound along it for this. It can be set to a gradation marked on it, but this is not as good as it might be as the thickness of the mark itself can make a difference of a quarter of a degree. It is better to set it against a known taper, using this as a pattern. To do this (assuming that we are turning a morse taper) put a centre mark in a short piece of steel bar mounted in the chuck and drill the bar with a Slocombe (centre) drill. Put the point of the lathe tailstock centre in this and then wind in the tailstock so that the point supports the other bar against the other centre. Now turn over the top-slide and, using a clock gauge or similar item, run it up and down the centre until the gauge shows the same reading all the way along. If you do not have a gauge, then a scriber or even a lathe tool will do, but in this case a piece of white paper should be held underneath to show up any variation between the angle of the top-slide and the taper. If it is not a morse taper being turned, that is a non-standard taper, then cut a triangle of card with the required angle on one side. Put the square end against the faceplate and adjust the top-slide in the same manner as before. The card edge must of course be straight.

Longer tapers will have to be turned by setting the work over at the tailstock end. Most tailstocks will move over and the work can be supported between centres. I must confess that I do not like setting the tailstock over and fight shy of doing so, as it can be extremely difficult to get it back again. One way round this is to make up a piece of flat metal with a bar to go in the tailstock chuck. Put in a small countersink at the required point and use a ball bearing as a centre. With careful measuring, this will do the job quite well, but may not be as accurate as you would wish. I have a slide attachment that does the job quite

Above *Setting a morse taper between centres in preparation for setting over the topslide.*

Below *Using a clock gauge to adjust the top slide for a taper. The clock gauge is like a clock with a spring pointer coming from it. As the pointer expands or retracts the hand on the clock gives a reading. To set a taper, the pointer must run along the known taper when the topslide is wound along and the hand should remain still. When there is no movement whatever the topslide is correctly adjusted.*

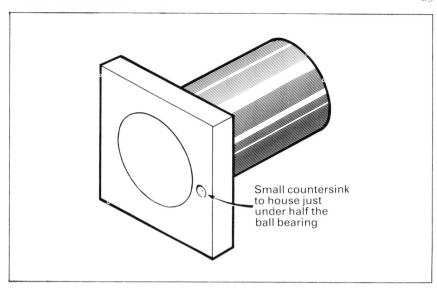

Small countersink
to house just
under half the
ball bearing

Above *Drawing of a metal block fixed to a round bar. The round bar is fixed in the tailstock chuck and the work which is to be held between centres is rested on a ball-bearing, which, in turn, fits into a small countersink in the metal plate. The countersink being off centre the work will run out of true at one end and a taper will be turned when a lathe tool is run along it. The offset countersink must be exactly half the amount required for the taper. For slight tapers the countersink can be put straight into the round bar.*

Below *A ball-bearing in use as a centre, on this occasion for normal parallel turning. If the ball was to be offset then a taper would be the result.*

Taper turning with the aid of a taper attachment that causes the cross-slide to move over as it is wound along the work. Note the uses of a steady and the revolving centre (Photograph courtesy of Myford Ltd.)

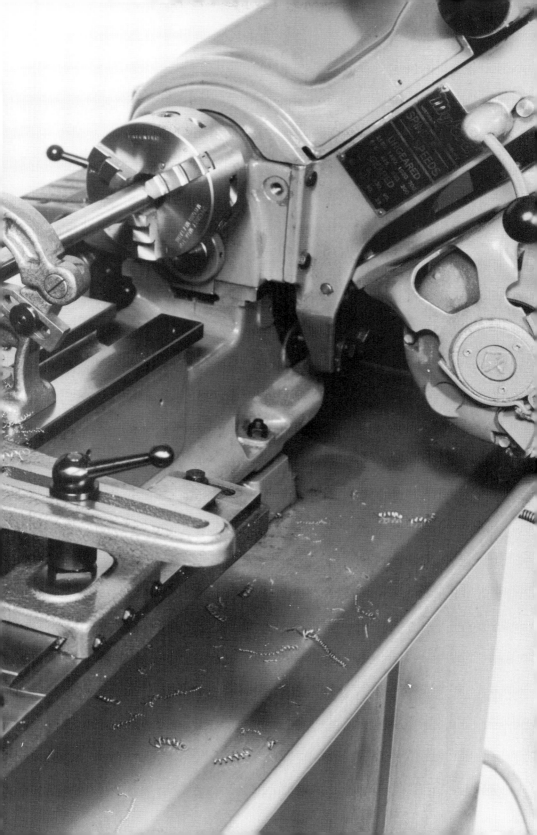

nicely and saves a great deal of bother. It is possible to purchase special attachments for taper turning and if a great deal of such work is to be done then it is worth doing so (or making one). They consist of a means of releasing the cross-slide from the lead-screw and a guiding bar along which, as the carriage is traversed along its bed, the cross-slide moves over.

REPETITION WORK

For small-batch production, such as hand-rail stanchions and similar items, some form of stop will be required for the carriage and the cross-slide. A simple, adjustable bar clamped on in some way so that it prevents further movement will do and even a piece of wood may suffice. If several tools are to be used then you will need some means of keeping them on the lathe without taking them off, as otherwise resetting up of the lathe would be necessary. A combination of rear tool-post and four-position tool-post can often prove useful. A tool-post turret can be made up for the smaller lathes, using a large nut that has holes drilled in it to take the various tools, which are held in by screws. Turrets can be also purchased, which fit either into the tailstock or on the cross-slide.

Below *A simplified drawing of a metal bar with a bolt in it, in use as a stop. The saddle strikes the bolt and because the block rests against the headstock it cannot travel any further. The bolt makes the arrangement adjustable. For small-batch production, a number of blocks can be interchanged. Pieces of wood cut to length will also do the job.*
Above right *A turret for use on the cross-slide which allows several tools to be held at once.*
Below right *A turret for use in the tailstock; this also allows the use of several tools at once.*

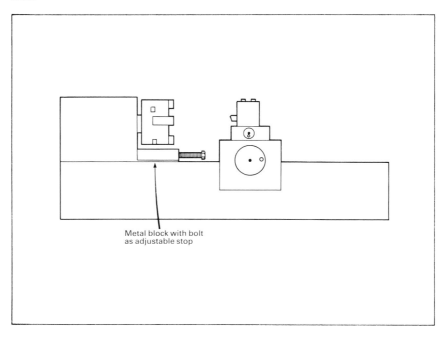

Metal block with bolt
as adjustable stop

KNURLING

Knurling is used to provide a grip on hand tools and to improve their appearance. Knurling is usually either of a straight or diamond pattern. A straight knurling tool is easily and cheaply made using a square bar of steel with a slot cut in it for the knurling wheel and a hole for the spindle on which the wheel will revolve. Cigarette-lighter wheels make good knurling wheels and are very cheap to buy. However, a brass or bronze bush must be put through the hole in the wheel before the spindle as, being so hard, the wheel can cut through a steel spindle in a matter of minutes. Diamond knurling is carried out with two wheels which are brought together at the top and bottom of the work whilst it is rotating. Some form of screw-down arrangement will be required to get the pressure on the work and a tool for this purpose, of the type shown, is easy and cheap to construct. When a knurling wheel has been set into the work, it should be wound along very slowly, keeping the pressure on. Cutting oil or lubricant must be used when knurling.

Left *A simple knurling tool made from a cigarette lighter wheel. It fits in a slot in a steel bar and has a hardened steel pin through it, which allows it to revolve on the work.*

Below *A home-made caliper-type knurling tool. The wheels are brought to bear on either side of the work by tightening up the wing nut on the top. A spring will open them when the wing nut is released.*

DIVIDING

Sooner or later, the need will arise to obtain exact divisions of the circumference of a piece of metal, an operation known, as one might expect, as dividing. The sort of thing I have in mind is putting square ends on spindles, making nuts and bolts of a special size and similar such jobs. In particular dividing is likely to be needed when milling operations are being carried out, as, for instance, when milling a square end to an axle. It is easy enough to measure out divisions with scibers and dividers but I wonder how accurate this is likely to be. Some far more positive form of division is required. Dividing devices can be purchased and they are excellent. They are also very expensive, when one considers the limited use to which they will be put. Most of our division work will be done when the job is in the chuck, so first let us think about using this for dividing. By using the chuck jaws we have ready-made divisions of three and four, assuming we are the proud owners of both forms of chucks. In order to make sure that the position on the work is right we need a positive means of ensuring that each jaw can be stopped in the same place. This is actually fairly simple; all we need is a piece of metal of suitable length for the jaw to rest on at ninety degrees and, each time we turn the chuck round to the next jaw, we know it will stop in the same position. To get divisions of six or eight we need a piece of metal of exactly half the length of the first one and so on. As usual there is a snag—the need to hold the lathe mandrel still whilst work is being carried out. If the key is put in the chuck, a piece of string can be tied firmly round the key and attached to the lathe

Using a block of wood under a chuck jaw for dividing. A 12-volt drill is mounted on the cross-slide for drilling a brass fitting.

stand. Providing it is pulled hard on to the metal strip it should cause no problems. The purists will shudder at the idea but it works. Make sure the lathe cannot be switched on accidentally while you are doing this, as by now, no doubt, like the rest of us, you will have at some time switched on the lathe with the key in the chuck. It is quite incredible how far the key can travel! It is also quite surprising how much the eyes water if it happens to hit you on the nose as it departs from the chuck!

The lathe 'change' wheels can be use for dividing and all that is needed is some sort of positive means of location. There have been numerous devices designed for this purpose but, as this book does not pretend to be more than just an introduction to lathe work, describing them here is not possible. The change wheels can also be pressed into service for work clamped to the cross-slide, whilst using a cutter rotating in the lathe. Again, all you need is a simple clamping device, so that the work and a change wheel with a suitable number of teeth can be rotated together when required, and some form of locking device on the wheel.

This drawing shows a simple, home-made filing rest. It is suitable for the majority of work, but if high precision is needed, then it will need a finer form of adjustment.

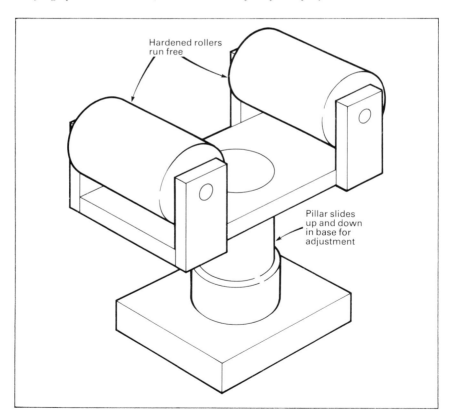

Hardened rollers
run free

Pillar slides
up and down
in base for
adjustment

FILING

Frequently when putting flats on a piece of metal held in the chuck, it is easier to use a file than to set up a milling attachment. It is very difficult to keep the file flat and if you are trying to file a square on the end of a round rod the result is, more often than not, an absolute disaster. What is needed is a rest that will ensure that the file remains flat and stays level. This, combined with our simple dividing device, will do the job as well as milling, if care is taken. Once more, simple wooden blocks can be pressed into service if nothing else is available. A piece of wood of identical thickness on either side will keep the file quite flat, even if it is a trifle inconvenient, A simple file rest is easy enough to make. It consists of two rollers, one on either side of the work, which should be hardened, but not tempered. They are fitted into a simple frame that will clamp to the cross-slide. If a height adjustment can be incorporated so much the better. It need not be elaborate, so long as it does the job. The sketch will give some idea of how such a simple device can be made.

I hope that what I have written will be of some help to the beginner in lathe work. There is a great deal more that could be written, but much can be discovered for oneself. Remember to try the easy way first, it will often work and save a great deal of effort. However, whether you are trying an easy way or not, care is important. Guesswork will not do and will only waste effort and metal. Measure first, check your measurements and then, when you are absolutely certain you are right, check once more. Above all enjoy yourself.

APPENDIX A

SAFETY

The importance of safety when dealing with a lathe cannot be over emphasised. The most obvious rule is to turn off and disconnect the lathe before changing belts, gear wheels etc. Chuck keys should never be left in chucks. Loose sleeves are dangerous, as is long hair. Both are likely to be caught in revolving machinery and the results can be at the least very painful. Ties also are dangerous and should not be worn. Any cutting fluid spilt on the floor should be wiped up at once. It is invariably based on oil or soap or both and is very slippery. It is easy to slip on it and to collide with moving machinery or sharp tools. In industry, guards over chucks are compulsory, while at home they are seldom used. If you do not have a guard, make sure that safety glasses are worn, as a steel splinter in the eye could mean blindness for life. Some cutting fluids are likely to upset some people. If after using one you find you are coughing excessively try a different one or wear a mask. They can also cause skin rashes. Wash thoroughly after using them and if there are any problems change to another brand. A good barrier cream can help to prevent skin rashes. Make sure that work is securely held in the lathe so that it does not fly out.

APPENDIX B

'DO'S' AND 'DON'TS'

Always make sure the tool you use is sharp. Do not be tempted to think as it was only used once it will do again. Sharp tools are essential for good turning. The only exception is when drilling brass or copper and then the edge should be taken off the drill with a small oilstone. Do get as near to the correct cutting speed as possible, the wrong speed means a bad finish. Always use the correct cutting fluid, but do not use anything on cast iron, as it will harden and become unworkable.

Do put tools away as you go, a cluttered lathe is an invitation to bad workmanship. Taps and drills should be kept in stands. If they are put in boxes or tins they will rub together and lose their edges. The same applies to reamers and end mills. Don't drop heavy objects on the bed of the lathe. They will at the least mark it and could cause permanent damage.

Do clean the lathe regularly. Remove the chuck and clean out where it fits to the mandrel. Take out the jaws and clean both those and the scrolls. Remove slides and clean them from time to time. Most lathes have strips in the slides called jib strips. These will have an adjustment on them, usually a screw and locknut. They should be adjusted so that there is no side play though excessive pressure is not required to move them. Do check the lathe mountings from time to time. A lot of pressure is put on them and they are likely to work loose. When not using the lathe, do cover all bright parts in a thin oil or one of the special anti-rust sprays, such as WD40.

APPENDIX C

WHAT CAN BE MADE?

What to make will largely depend on individual interests. Apart from ornamental turning, a certain amount of bench work will be required for most things. Here is a list of possibilities:

Model stationary steam engines.
Model locomotives.
Model traction engines.
Model cars and lorries.
Model internal combustion engines.
Hot air engines.
Model cannons.
Parts for model boats and aircraft such as hand-rail stanchions, propeller couplings etc.
Candlesticks.
Small metal boxes.
Fishing reels.
Telescopes,
Tools.
Door knobs.
Parts for doll's house furniture.

INDEX

Adjustable tool posts, *50*
Angle plates, *43-80*

Back gears, *10-14*
Bed, *8, 14*
Benches, *21*
Blind bores, *65*
Boring bars, *61*
Boring between centres, *62*

Carbide tips, *29*
Carbon steel, *27*
Carriage, *10*
Catch plates, *38*
Centre drills, *39, 51*
Centre height, *8*
Centres, *23*
Change wheels, *90*
Chasers, *72*
Checking alignment, *22*
Chucks, *14*
Clock gauge, *81*
Collet chucks, *46*
Cross slide, *10*
Compound slide, *8*
Cutting angles, *26*
Cutting lubricants, *21, 53, 57*
Cutting speeds, *54, 58*
Cutting tools, *26*

D bits, *59*
Die holders, *68*
Dividing, *89*
Drills, *57, 59*

Drive pulleys, *8*
Driving dogs, *38*

Face plates, *8, 47*
Filing rest, *90*
Fly cutter, *76*

Head stock, *8, 10*
Hardening, *30*
High speed steel, *27*

Integral motors, *10*

Knurling, *88*

Lathe carriers, *40*
Lead screw, *8*
Levels, *22*

Mandrel, *8*
Milling cutters, *75*
Morse tapers, *10, 60*

Parting off, *33*
Parting tools, *33*
Pin chucks, *58*

Rear tool posts, *12, 51*

Saddle, *8*
Screwcutting gears, *71*
Second-hand lathes, *14*
Shims, *48*
Steadies, *53*
Stops, *82*

Tailstock, *8*
Tailstock chuck, *12*

Tailstock die holders, *12*
Tapers, *81*
Tapping device, *67*
Tee slots, *10*
Tempering, *30*
Threading, *68*
Thread finishing, *72*
Thread gauge, *70*
Thread dial indicator, *71*
Thread shapes, *73*
Tool height, *24*

Tool posts, *8, 48*
Tool posts: four-way, *49*
Tool Posts: home-made, *51*
Tool posts: improved, *51*
Top slide, *8*
Turning, *53*
Turrets, *83*

Vee blocks, *80*
Vernier calipers, *60*
Vertical slides, *77*